S0-BLA-257

Regional residuals environmental quality management modeling

BLAIR T. BOWER, Editor

RESEARCH PAPER R-7

RESOURCES FOR THE FUTURE / WASHINGTON, D.C.

RFF Research Papers

R-1. *Environmental Quality Management: An Application to the Lower Delaware Valley.* Walter O. Spofford, Jr., Clifford S. Russell, and Robert A. Kelly. 1976.

R-2. *Public Regulation of Site-Selection for Nuclear Power Plants: Present Procedures and Reform Proposals—An Annotated Bibliography.* Ernest D. Klema and Robert L. West. 1977.

R-3. *Research in Forest Economics and Forest Policy.* Marion Clawson, editor. 1977.

R-4. *Decision Making in Timber Production, Harvest, and Marketing.* Marion Clawson, editor. 1977.

R-5. *Modeling Energy–Economy Interactions: Five Approaches.* Charles J. Hitch, editor. 1977.

R-6. *The World Food Situation: Resource and Environmental Issues in the Developing Countries and the United States.* Pierre R. Crosson and Kenneth D. Frederick. 1977.

R-7. *Regional Residuals–Environmental Quality Management Modeling.* Blair T. Bower, editor. 1977.

RFF Working Papers

EN-1. *Energy Modeling: Art, Science, Practice.* Milton F. Searl, editor. 1973.

EN-2. *Patterns of Energy Consumption in the Greater New York City Area: A Statistical Compendium.* A joint study by the Regional Plan Association, New York City, and Resources for the Future. 1973. Out of print.

EN-3. *Energy and the Social Sciences: An Examination of Research Needs.* Hans H. Landsberg et al. 1974.

EN-4. *U.S. Energy R&D Policy: The Role of Economics.* John E. Tilton. 1974.

EN-5. *Mineral Materials Modeling: A State-of-the-Art Review.* William A. Vogely, editor. 1975.

EN-6. *The Economics of National Forest Management.* Marion Clawson. 1976.

LW-1. *Forest Policy for the Future: Conflict, Compromise, Consensus.* Marion Clawson, editor. 1974.

PD-1. *Changing Resource Problems of the Fourth World.* Ronald G. Ridker, editor. 1976.

PD-2. *The Politics of Environmental Reform: Controlling Kentucky Strip Mining.* Marc Karnis Landy. 1976.

QE-1. *Ecological Modeling in a Resource Management Framework.* Clifford S. Russell, editor. 1975.

Research Papers are studies and conference reports published by Resources for the Future from the authors' typescripts, without the usual editorial review. The Research Paper series is intended to provide inexpensive and prompt distribution of research that is likely to have a shorter shelf life or to reach a much smaller audience than RFF books. This series replaces the Working Paper series.

Regional residuals environmental quality management modeling

BLAIR T. BOWER, Editor

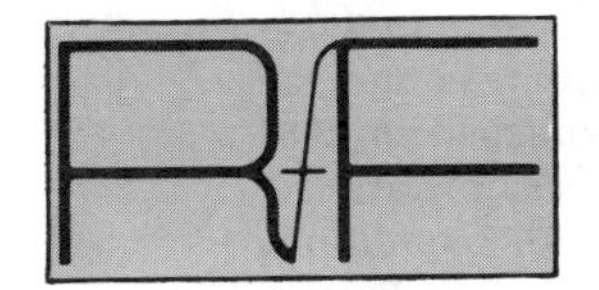

RESEARCH PAPER R-7

RESOURCES FOR THE FUTURE / WASHINGTON, D.C.

Resources for the Future is a nonprofit organization for research and education in the development, conservation, and use of natural resources and the improvement of the quality of the environment. It was established in 1952 with the cooperation of the Ford Foundation. Grants for research are accepted from government and private sources only if they meet the conditions of a policy established by the Board of Directors of Resources for the Future. The policy states that RFF shall be solely responsible for the conduct of the research and free to make the research results available to the public. Part of the work of Resources for the Future is carried out by its resident staff; part is supported by grants to universities and other nonprofit organizations. Unless otherwise stated, interpretations and conclusions in RFF publications are those of the authors; the organization takes responsibility for the selection of significant subjects for study, the competence of the researchers, and their freedom of inquiry.

Library of Congress Catalog Card Number 77-92413
ISBN 0-8018-2096-0

Manufactured in the United States of America

Published November 1977. $6.50.

TABLE OF CONTENTS

LIST OF TABLES

LIST OF TABLES (continued)

LIST OF FIGURES

LIST OF FIGURES (continued)

AFFILIATIONS OF AUTHORS

At time of Rotterdam meeting	Present
D. J. Basta	
Johns Hopkins University/Urbanistični Inštitut SRS Jamova 18 61001 Ljubljana, Yugoslavia	Resources for the Future 1755 Massachusetts Avenue, N.W. Washington, D.C. 20036
B. T. Bower	
Resources for the Future	Same
C. W. Howe	
Department of Economics University of Colorado Boulder, Colorado 80203	Same
R. A. Kelly	
Resources for the Future	Fisheries and Wildlife Department 605 Flinders Street Extension Melbourne, Victoria 3000, Australia
A. V. Kneese	
Department of Economics University of New Mexico Albuquerque, New Mexico 87131	Same
J. L. Lounsbury	
Johns Hopkins University/Urbanistični Inštitute SRS	Office of Management and Budget Natural Resources Division, Environment Branch New Executive Office Building Washington, D.C. 20503
B. Reznicek	
Czechoslovak Research and Development Centre for Environmental Control Srobranova 48 10042 Prague 10, Czechoslovakia	Same
C. S. Russell	
Resources for the Future	Same
W. O. Spofford, Jr.	
Resources for the Future	Same

To
JOACHIM KUMPF

An exemplary international civil servant

PREFACE AND ACKNOWLEDGMENTS

This report had its genesis at a meeting on modelling for regional residuals-environmental quality management held in Rotterdam, The Netherlands, 22-25 October 1974. The meeting was jointly sponsored by the European Regional Office of the World Health Organization, Copenhagen, and Resources for the Future, Washington, D.C., in collaboration with the Government of The Netherlands. Arrangements for the excellent physical facilities and associated inputs for the meeting in Rotterdam were made by Dr. K. Biersteker of the Municipal Service for Medicine and Public Health of Rotterdam and were made available through the courtesy of the Municipal Service. The Quality of the Environment Division of Resources for the Future financed the preparation and publication of the report.

The meeting would not have taken place without the interest, encouragement, and support of Joachim Kumpf. During the time he was Chief, Protection of Environmental Health, European Regional Office of WHO, he led the way toward incorporating an integral view of the three environmental media in the work under his direction. Hence, it is fitting that this report be dedicated to him.

That the meeting in Rotterdam ran smoothly was a function of the skill of Mette Christiansen, at that time of the WHO staff in Copenhagen. At Resources for the Future, Margaret White, Patricia McHenry, and Luz Maria Aveleyra typed the various drafts and the final version of the report, and Pathana Thananart drew most of the figures.

The report itself is a composite of material developed during different periods of time. Chapter 2 was prepared prior to the Rotterdam meeting, to provide a starting base for the discussions. Chapter 7 was originally prepared shortly after the meeting and reflects the discussions at the meeting. Chapters 3 through 6 represent the regional case studies presented at the Rotterdam meeting, but with substantial additions made subsequent to the meeting to chapters 3 and 5 after the Lower Delaware Valley and Ljubljana studies, respectively, were completed. Chapter 1 was prepared to provide an introduction to the subject matter.

This research report is one of three closely related RFF publications. It provides an overview of regional residuals-environmental quality management and of the analyses necessary to generate information for decisions on such management. The two publications related to this report are elaborations of the case studies presented in chapters 3 and 5. _Environmental Quality Management: an Application to the Lower Delaware Valley_, published by RFF in 1976, is a detailed description of the case study for REQM of one of the most highly industrialized regions in the United States. _Analysis for Residuals-Environmental Quality Management: A Case Study of the Ljubljana Area of Yugoslavia_, to be published by RFF in 1977, is a detailed description of the case study for REQM of a metropolitan area in a rapidly developing country with a political-economic system substantially different from that of the United States. The Lower Delaware Valley and Ljubljana area studies illustrate the application of different analytical approaches to analysis for REQM. Together, the three publications provide background for, and insights on, the analytical problems faced in developing useful information for societal decisions on residuals-environmental quality management.

Blair T. Bower
September 1977

Chapter 1

THE WHY AND WHAT OF REGIONAL RESIDUALS - ENVIRONMENTAL QUALITY MANAGEMENT MODELLING

Blair T. Bower

Introduction

As awareness of the adverse impacts of man's activities on the "natural" environment has increased, so has cognizance increased that positive action must be taken if ambient environmental quality (AEQ) is to be improved and maintained. AEQ reflects the quality of the water, air, and land environments as measured by such indicators as: concentration of dissolved oxygen in a river or lake; biomass of fish per unit volume of a water body; concentration of nitrates in ground water; concentration of sulphur dioxide in the urban atmosphere; hectares of land disturbed by surface mining; and hectares of land having uncontrolled dumps of solid residuals. Decisions must be made in all societies concerning: what levels of AEQ to seek; what the costs are of achieving different levels of AEQ by different means; who pays the costs and who benefits from the improved AEQ; and what implementation incentives and governmental institutions are necessary to induce the desired responses by the activities which impose "loads" on the environment. Such decisions are made in any society within the context of a set of social, economic, political, technological, and informational constraints.

Whatever the societal and institutional context, making decisions on how to improve AEQ requires--or should require--information on the costs and consequences of alternative strategies. The rigorous development of such information requires analysis, such a requirement being the basic rationale for regional residuals-environmental quality management modelling.

Key Definitions and Concepts

A few terms need to be defined and a few concepts explained to provide a common foundation for the subsequent chapters.

Residuals

All human activities--households, farming, manufacturing, mining, transportation--result in the generation of residuals. This is because no production or use activity transforms all of the inputs to the activity into desired products or services. The remaining flows of materials and/or energy from the activity are termed nonproduct outputs. If a nonproduct output has no value in existing markets or a value less than the costs of collecting, processing, and transporting it for input into the same or another activity, the nonproduct output is termed a residual. Thus, residual is defined in an economic sense. Hence, whether or not a nonproduct output is a residual depends on the relative costs of alternative materials and/or energy which can be used instead of the non-product output. These costs in turn depend on the level of technology in the society at the point in time and on various governmental policies, both of which can change over time.

Figure 1-1 illustrates the definition of residual, by means of a simplified flow diagram of the production of paper napkins. Some of the nonproduct outputs generated are directly recovered and reused in the production process, such as chemicals from the pulping process and fiber from the paper machine. Depending on relative costs of the alternative factor inputs, more or less recovery and reuse of these non-product outputs will be undertaken in the absence of constraints of one type or another on discharges of residuals to the environment. That is, materials and energy recovery will, in in principle, take place up to the level where the marginal cost of an

Figure 1-1. Illustrating the Definition of Residuals: Simplified Process Flow Chart for Production of Paper Napkins

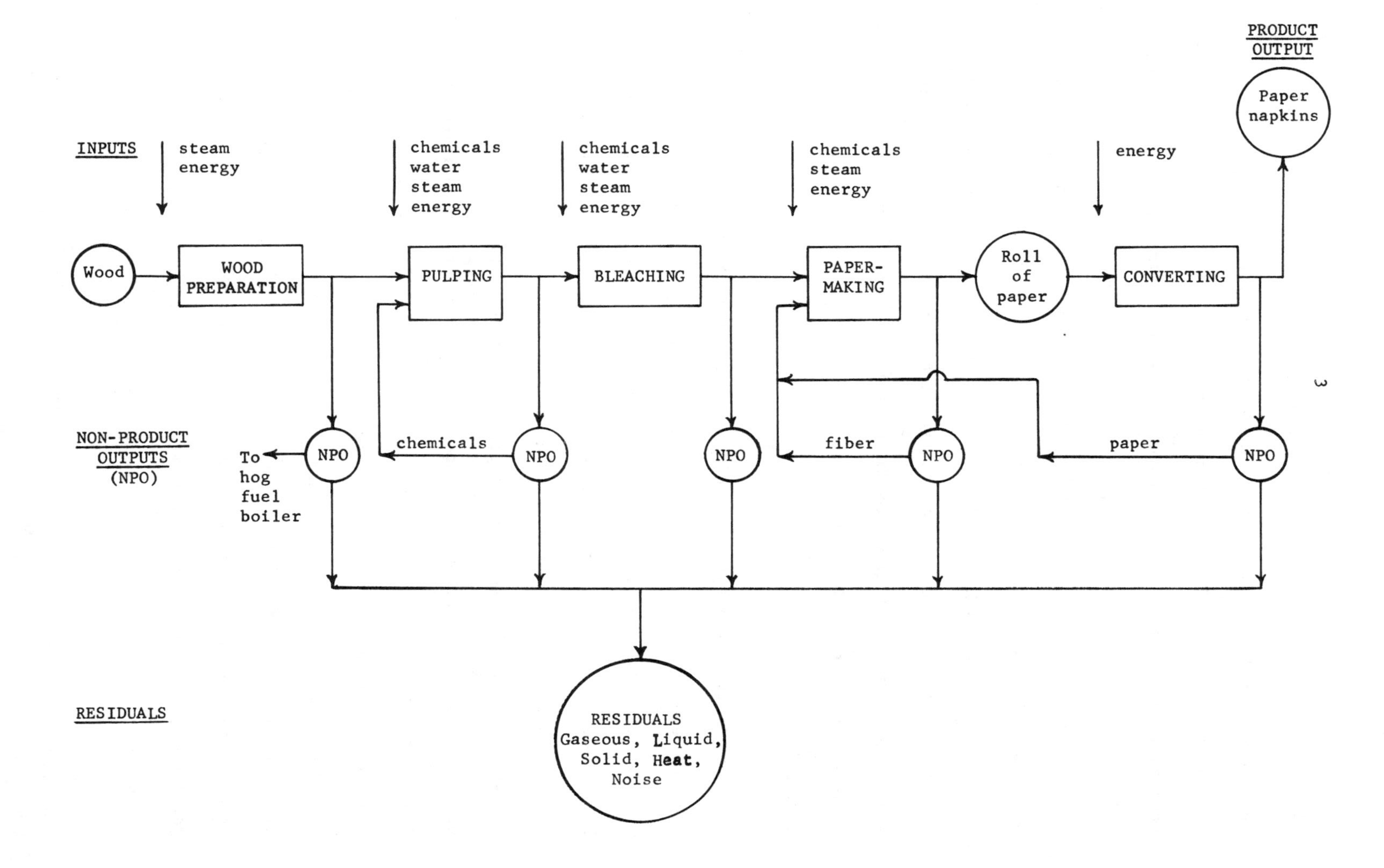

additional unit of recovered material or energy just equals the value of the recovered material or energy, as determined by the pricing system in the society. Because both the cost of recovery and the value of the recovered material or energy often change over time, the extent of recovery, and hence residuals generation, are likely to vary over time. The remaining nonproduct outputs--the quantities remaining after economical recovery has been undertaken--are residuals.

Material and energy are the two basic classes of nonproduct outputs and residuals. The former type occurs in the three states of matter--liquid, gaseous, and solid. The major energy residuals are heat and noise. Radioactive residuals have characteristics of both material and energy residuals.

Interrelationships Among Residuals. It is critical to recognize and consider explicitly the interrelationships among residuals. One form of material residual can be transformed into one or more other forms, usually by the addition of materials and energy, as in the modification of sewage in a municipal treatment plant. The modification requires materials and energy inputs which also become residuals. Modification is presumably undertaken under the assumption that the discharge of the modified residual and the residuals generated in its modification will have less adverse impacts than the discharge of the original residual.

These interrelationships can be illustrated by considering a power plant using coal as the fuel for electric energy generation. The particulates formed in combustion can be discharged to the atmosphere in the exhaust gas stream, i.e., a gaseous residual. If, however, there are constraints on such discharge a wet scrubber could be installed to wash the particulates out of the gas stream, thereby transforming the gaseous residual into a liquid residual, i.e., suspended solids, which could then be discharged to an adjacent water body. Such discharge might adversely affect water quality, with consequent damage to fish. To prevent such an impact, a settling basin could be installed to settle out the suspended solids in the liquid residual, thereby yielding a "solid" residual for ultimate disposal.

With respect to energy residuals, approximately two kilowatt hours (kwh) equivalent of residual heat are generated by the power plant for every kwh of desired energy output (the "product") produced. This residual (waste) heat is discharged to the environment, traditionally by direct discharge to water courses, i.e., once-through cooling. When there are adverse effects from such discharge, a typical response is to install a cooling tower, which results in the discharge of the thermal residual to the atmosphere first instead of directly to the water. Such discharge may in turn have adverse effects, such as increased fog in the area, icing in winter, and increased precipitation in the immediate area.

Factors Affecting Residuals Generation

Having defined residuals, the factors affecting residuals generation can be delineated. That is, what factors affect the types and quantities of residuals generated in various activities--manufacturing, mining, agriculture, households--per unit of activity, such as per ton of product, per barrel of crude petroleum processed, per bushel of wheat, per capita, per household? In the absence of controls on discharges of residuals, the residuals generated in manufacturing, for example, are a function of: the characteristics of the raw materials used; the technology of the production process--including age and physical arrangement of plant; the product mix;

the specifications of each of the desired products; the operating rate--units of raw material processed or units of output per unit of time; and prices of factor inputs. Referring to Figure 1-1, the residuals generated per ton of paper napkins produced are a function of: the species of wood used; the method of wood preparation; the type of pulping process; the characteristics desired in the final product, such as wet strength, softness, absorbency, and color; the type of paper machine; the number of tons of paper produced per hour; and the prices of various factor inputs, such as fuel, chemicals, wood, water, electric energy. The degree of whiteness desired is of particular importance, because this determines the amount of bleaching required and bleaching is a major source of residuals generation in the production of paper products. Shifting from white to unbleached paper products wherever possible, i.e., brown instead of white napkins, while holding all other product specifications constant, would substantially reduce residuals generation.

Similarly, the residuals generated in producing wheat are a function of such factors as: the type of soil; topography; type of cultivation equipment; cultivation practices; frequency and types of fertilizer application; frequency and types of pesticide application; prices of factor inputs; and climate. With respect to households, residuals generation is a function of: design of structure; exposure; whether a single unit or multi-unit structure; size of yard; income; education; climate; and prices of water, electric energy, fuel, sewage disposal, solid residuals disposal.

The dynamic character of residuals generation in any activity merits emphasis. Changing unit generation results from both endogenous and exogenous factors. Typically, for a given type of activity--production of paper or steel or wheat--residuals generation per unit changes over time as a result of changes in endogenous factors such as: production technology; types of raw materials used; product mix; and product specifications. Exogenous factors affect residuals generation in an activity both directly and indirectly. Changes in prices of such factor inputs as fuel, water, electric energy, chemicals, labor will usually result in changes in residuals generation by a given activity. Similarly, the value of a given material or energy nonproduct output generated by an activity is very likely to change over time in relation to the value of competing materials and energy for a particular use. An example is copper wire as a secondary (nonvirgin) source of metallic copper in relation to copper ore as a source of copper. As high quality copper ore deposits have been exhausted and the technology of processing copper wire to produce copper has become more efficient, the value of copper wire relative to copper ore has increased.

REQM System

The concept of a residuals-environmental quality management (REQM) system is illustrated in figure 1-2. Within any given region (however defined) at a given point in time, there is a spatial distribution of activities: industrial, mining, residential, agricultural, commercial, transportation. This spatial distribution of activities reflects the final demand for goods and services within the region and from outside the region. For each activity there are: (1) alternative combinations of factor inputs and related production technologies to produce the required goods and services, with a set of types and quantities of residuals generated associated with each combination; and (2) alternative ways of handling

Figure 1-2. Concept of a Residuals-Environmental Quality Management (REQM) System

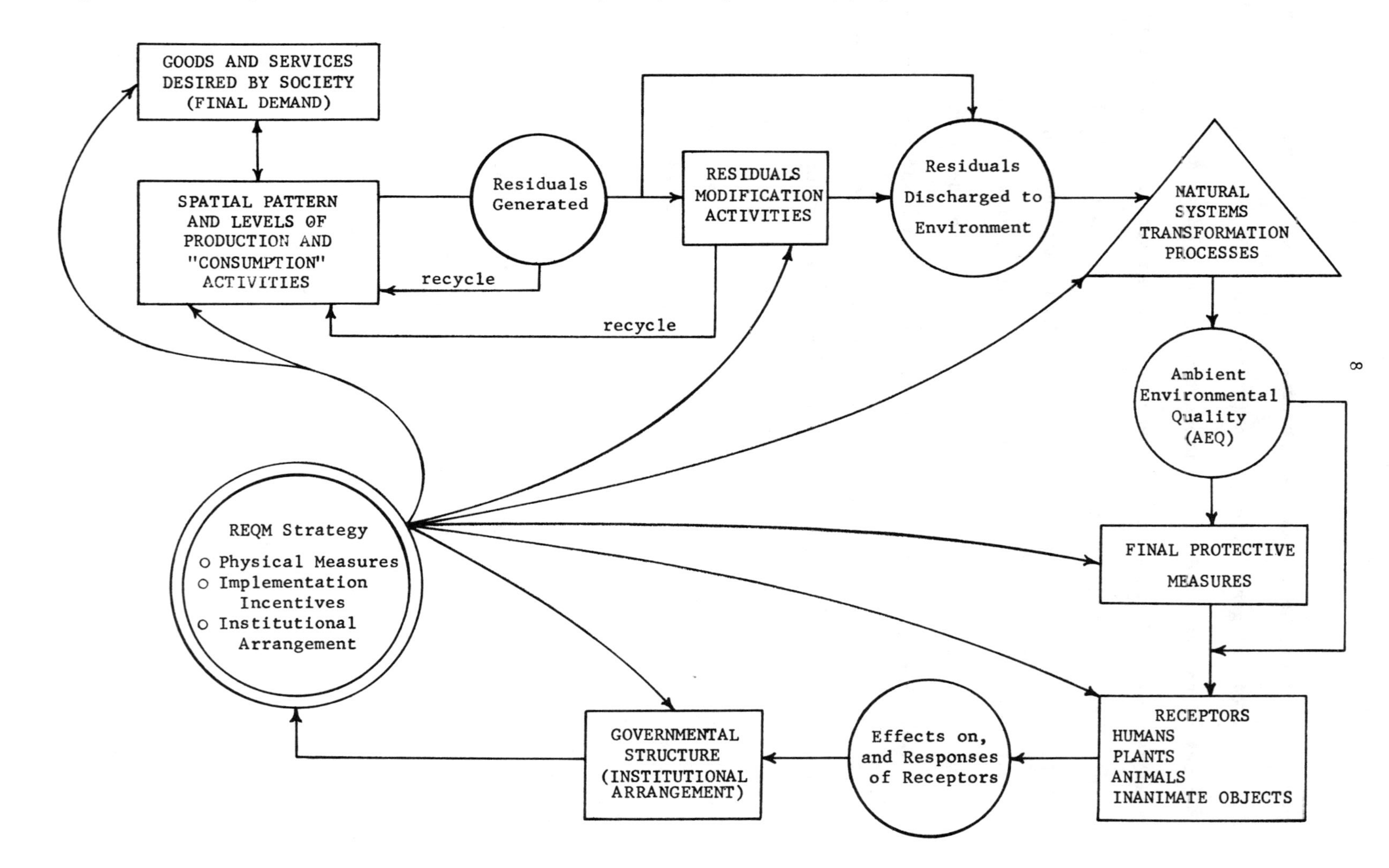

the residuals after generation. Activities can be characterized as point (manufacturing plant, residence), line (traffic flow on major street), or dispersed area (logging, agricultural crop production) sources of residuals.

From each activity some residuals are directly or indirectly discharged into the air, water and/or land environments. In the environment these residuals are affected by and may affect various physical, chemical, and biological processes--transport, sedimentation, absorption, adsorption, volatilization, decomposition, accumulation. These processes transform the time and spatial pattern of residuals discharges from the various activities into the resulting short-run and long-run time and spatial patterns of AEQ. What indicators are used at any point in time is a function of: (1) the existing knowledge about the effects of the residuals represented by the indicators; (2) the ability to measure the indicators, i.e., the technology of measurement; and (3) the available data.[1*]

The resulting time and spatial pattern of AEQ impinges directly on the receptors--humans, plants, animals, materials such as structures--or indirectly, as where "final protective measures" are installed between the ambient environment and the receptors. Final protective measures are exemplified by a water intake treatment facility which modifies the quality of water withdrawn from a water body before the water is distributed for use. The impacts on the receptors, as perceived by human beings, and the responses of individuals and groups to the perceived damages, provide the stimulus for action. The extent and form of action, as expressed in a selected REQM strategy, depends on the institutional structure, culture and value system, and competing demands for scarce resources for other desired goods and services.

---

*Footnotes are at the end of each chapter.

Residuals-Environmental Quality Management (REQM)

REQM consists of the following functions required to produce a desired level of AEQ: analysis to develop REQM strategies; planning; legislation; translation of legislation into guidelines and procedures; implementation of guidelines and procedures via incentives imposed on residuals generating activities to induce those activities to install and operate physical measures for reducing the discharge of residuals into the environment and/or for modifying or making better use of the available assimilative capacity; design/construction/operation of facilities; monitoring and enforcement of performances by activities; monitoring of AEQ; and feedback of information from monitoring into the continuous planning and decision-making functions. It is this total set of functions which yields the desired product of improved AEQ.

REQM on a day-to-day basis takes place primarily at the regional level, ergo, regional REQM. The region may be a metropolitan area, a river basin, an airshed, a soil conservation district, an economic region, or some combination of local jurisdictions. Experience indicates that it is less important that the boundaries of the region include all of the residuals dischargers and all those affected by changes in AEQ, than it is that the boundaries represent some region or area for which there is an institutional arrangement which can be made responsible for REQM. However, where cross-boundary transfers of residuals and their effects are significant, they must be explicitly considered. It is also extremely important to recognize that factors such as national tax policies and the prices of factor inputs established in national markets affect REQM in a region. Such factors are determined exogenously to the region, and comprise part of the context in which REQM takes place and are major factors which should be--but often have not been--explicitly considered.

REQM Strategy

An REQM strategy is comprised of: (1) the physical measures for improving AEQ; (2) the implementation incentives to induce the residuals generators to apply the measures; and (3) the institutional arrangements through which the implementation incentives are applied and the other related activities of REQM are carried out.

Physical Measures for Improving AEQ. The two basic classes of physical measures for improving AEQ are shown in table 1-1: (1) reducing discharge of residuals to the environment; and (2) making better use of, or increasing the assimilative capacity of, the environment. The first class has two subclasses: (a) reducing generation of residuals; and (b) modifying residuals after generation. Reduction in generation can be accomplished by changing the characteristics of goods and services desired, changing production processes, and changing raw materials. Often these three are interrelated, that is, a change in product specifications will enable or require changes in both production processes and raw materials. For a residence, changing plumbing fixtures and appliances represents analogous changes to changes in technology of production in a manufacturing plant.

It is important to recognize that physical measures can be adopted by a given activity on-site which will have no effect on residuals generation by the activity itself. For example, physical measures which result in reduced electrical energy used by an activity will reduce the quantity of kilowatt hours which have to be generated and hence the associated residuals generated in producing the electric energy. Similarly, if the management

Table 1-1. Physical Measures for Improving Ambient Environmental Quality

Measures for Reducing the Discharge of Residuals

Measures for reducing residuals generation

1. Change raw material inputs
2. Change production processes
3. Change mix of product outputs
4. Change product output specifications
5. In-plant (on-site) recirculation of water [a]

Measures for modifying residuals after generation
(May take place in single or joint facilities)

1. Materials or energy recovery (direct recycle)
2. By-product production (indirect recycle)
3. Residuals modification without recovery for reuse of any material or energy
4. In-plant recirculation of water [b]

Measures Directly Involving Environmental Assimilative Capacity

Measures for making better use of assimilative capacity

1. Redistribute effluent from a given activity over space and/or over time [c]
2. Change the time scheduling of activities [d]
3. Change the spatial location of activities [d]

Measures for increasing assimilative capacity

1. Add dilution water to water bodies
2. Use multiple outlets from reservoirs
3. Artificially mix water in reservoirs and lakes
4. Artificially aerate streams, lakes, estuaries with air or oxygen, by surface or subsurface diffusers

---

a Generation of residuals, including waste water, changes per unit of activity

b No change occurs in residuals generation per unit of activity, except wastewater

c No change in the location, level, timing of the activity itself or in residuals generation by the activity

d Residuals generation per unit of activity may or may not change of a center-city office building shifts from bleached to unbleached paper towels--all other specifications of the towels remaining the same--residuals generation in the office building will not be changed but residuals generation at the loci of production of the towels will be. Of course, the reverse is also possible. Physical measures adopted by an activity on-site may not increase on-site residuals generation but may increase residuals generation elsewhere. The greater the linkages among activities in a given region--in terms of material and energy flows among those activities--the more important this fact becomes for REQM in the region.

Modification of residuals after generation can be accomplished by: materials recovery (direct reuse in the <u>same</u> production process); by-product production or indirect reuse (use of a residual as an input to a <u>different</u> production process, at the same or a different location); recirculation of water; and modification without recovery, traditionally termed "waste treatment". Except for materials recovery (as defined), these physical measures can be carried out in a facility handling residuals generated at multiple locations. Examples of joint facilities are: a waste oil reclamation plant which processes used oil from many garages and gas stations; a plant for recovery of acid from used steel mill pickling liquor from several mills; a municipal incinerator, landfill, sewage treatment plant. Joint facilities usually permit achieving economies of scale, thus making economically feasible an activity which would not be so if it were undertaken at each individual operation. As noted previously, residuals modification without materials and/or energy recovery does not reduce the total quantity of residuals discharged into the environment. Such modification merely transforms one type of residual into other forms of the same type and/or one or more other types. In order to make the transformation, additional inputs are required, and hence additional residuals are generated.

The second basic class of physical measures for improving ambient environmental quality also has two subclasses: (a) making better use of existing assimilative capacity; and (b) increasing assimilative capacity. The former reflects the fact that assimilative capacity varies both over space and time, e.g., diurnally, seasonally, year-to-year. Examples of the former are: (a) building stacks for discharging gaseous residuals at levels such that greater dispersion can be achieved to reduce ground level concentrations in the region[2]; (b) relocating outfalls to where there is greater assimilative capacity in streams and estuaries; (c) temporary withholding of liquid residuals during periods of low streamflow for subsequent discharge when streamflow, and hence assimilative capacity, is higher; (d) rescheduling of activities in time, such as changing the time pattern of traffic flow; and (e) relocating activities in space to be more in accord with the natural assimilative capacity in an area, such as locating industrial activities downwind from residential areas. (a), (b), and (c) involve no change in residuals generation. (d) and (e) may change residuals generation, as when rerouting of traffic results in increasing average velocity, which in turn means fewer residuals generated per vehicle mile traveled. Increasing the assimilative capacity can be accomplished by: the addition of water to streams during low flow periods; artificially adding oxygen to streams, lakes and estuaries; and building topography with solid residuals, i.e., making artifical hills.

Implementation Incentives. Implementation incentives are the inducements which stimulate the residuals generators and the REQM agencies to install, operate, and maintain the physical methods for improving AEQ. An implementation incentive can be positive or negative (reward or punishment), direct or indirect, prescriptive or proscriptive. Table 1-2 is a classification of implementation incentives. Although the categories and examples in this table are mostly self-evident, some further discussion is merited.

First, some of the implementation incentives can be imposed at more than one level of government. For example, an upper limit on the sulphur content of fuel has been specified by both local and state governments in the U.S. Similarly, restrictions on nonreturnable beverage containers have been imposed at both local and state levels. On the other hand, some incentives are more, or exclusively, relevant to one particular level of government. Requiring householders to keep used newspapers separate from other solid residuals is relevent to the local level; requiring automobiles produced in the U.S. to meet a minimum distance/fuel standard is relevant only at the federal level. Some incentives can be imposed incrementally at more than one level of government, such as a national effluent charge on suspended solids discharges into water courses combined with an additional state effluent charge on such discharges.

Second, mixes of implementation incentives can be applied to the same activity. An effluent standard in terms of mean daily kilograms of BOD_5 discharged can be coupled with an effluent charge ($/kg) on all BOD_5 discharges or on all discharges above a specified limit.

Third, incentives imposed on one residual can have positive and/or negative impacts on other residuals and/or environmental media. An ordinance specifying limits on concentration of particulate discharges from

Table 1-2: Classification of Implementation Incentives for Environmental Quality Management

Regulatory--by law, ordinance, permit

1. Specification of a physical method

   a. Specify characteristic(s) of raw material input, for example, no more than 1 percent sulphur fuel.

   b. Specify production "process," for example: dry peeling in fruit and vegetable canning; road design and construction in national forests with respect to grade, drainage, encroachment on stream channels; automatic turnoff valves on all water outlets in commercial, industrial, institutional facilities; orientation of buildings with respect to sun and wind; amounts of thermal and/or noise insulation in buildings; airplane movements on the ground; fencing, rotation, feeding and watering locations on grazing lands; operational procedures for landfills; frequency of street cleaning and litter removal in urban areas; design and construction standards for water and sewer pipes; restriction on type of pesticide and method of application; height of stacks; vehicle movement on certain streets.

   c. Specify residuals modification and/or handling process, for example: activated sludge; debris basins on construction sites; require householders to separate used newspapers from other solid residuals; prohibit acceptance of used newspapers/used corrugated containers at municipal incinerators.

   d. Specify product output characteristics, for example; no more than 8 percent phosphates in detergents; returnable beverage containers; number of sizes of cans for canned food; amount of lead in gasoline.

2. Specification of a result or performance

   a. Specify residuals discharge per unit of product or raw material processed $\leq$ specified amount, for example $\leq$ X kilograms of suspended solids (SS) per ton of steel or per barrel of crude throughput; $\leq$ Z grams of HC per vehicle-kilometer.

   b. Specify total quantity of a residual discharged per unit of time $\leq$ specified amount, for example: Y pounds of BOD_5 per day.

   c. Specify concentration of residuals in effluent $\leq$ specified magnitude, for example: $\leq$ 30 mg/1 of BOD_5, $\leq$ 30 mg/1 of SS.

   d. Specify AEQ should meet or exceed specified levels for specified periods of time, for example: > 6 mg/1 dissolved oxygen in spring and fall, $\leq$5 mg/1 remainder of year; mean annual concentration of sulphur dioxide $\leq$ 80 micrograms/m^3.

   e. Specify automobiles must achieve at least 40 kilometers per liter of fuel in city driving; appliance must last at least 8 years; residuals modification facility must achieve at least 85 percent removal of BOD_5 from a specified base; appliance must achieve at least a specified level of efficiency in energy use; building, plumbing, etc. codes based on performance over time, for example, heat loss specification.

   f. Specify producer must prove that a product is environmentally "benign," for example, pesticides.

Table 1-2 (Continued)

3. Specification of limitations on location of activity, for example: by zoning or land use regulations of various types, such as: restrictions on building where infiltration capacity for septic tanks is less than a specified rate; prohibition of strip mining on slopes < 25 percent with specified soil type; prohibition of building on slide-prone land; prohibition of development where utility capacity--water, sewer, slectricity--is insufficient.

4. Specification of extent, timing, type of activity, for example: prohibition of trucks on particular routes during particular times; prohibition of automobiles in central business district; prohibition of all-terrain vehicles in environmentally fragile areas; limit number of users (campers, boaters, horse riders) to $\leq$ a specified number per day; prohibit aerial spraying when wind $\geq$ 8 kilometers per hour; staggered work hours; reduction or cessation of production during adverse AEQ conditions.

5. Specification of procedure, for example: requirement that environmental impact statement for each project be prepared according to specified guidelines; requirement that planning for environmental quality management, for example, areawide water quality management (Section 208 of P.L. 92-500), be carried out according to certain procedures; registration of pesticides; excess packaging review; auto assembly line emissions testing; semiannual testing and inspection by state agency of emission controls on motor vehicles; requirement for public hearings.

Economic

1. Applied directly to residuals, for example: charge on each unit of residual discharged--cents per kilogram of BOD_5, cents per kilogram of sulphur, cents per calorie of heat (with surcharges based on timing of discharge); fines for spills or accidental discharges; sale of discharge rights.

2. Applied to inputs or product outputs, for example: charge on each kilogram of DDT applied; charge on each kilogram of phosphates in detergents; charge on packaging--cents per kilogram of plastic/paper used; surcharge on horsepower of automobile at time of purchase; annual surcharge on horsepower of automobile; severence taxes on virgin materials extracted; depletion allowances on virgin materials extracted; capital gains provisions for virgin materials extracted; subsidies for use of secondary materials; expensing provisions relating to extraction of virgin materials; water intake pricing; surcharge on energy use; differential taxes on fuels--$>$ gas, $<$ diesel, $<$ methanol; surcharge for power equipment in vehicles--air conditioners, power steering and power brakes, automatic transmission, except for vehicles used by disabled individuals.

3. Applied to activities, for example: reduced parking fees for car pools; parking surcharges; provision by agency or firm of multi-passenger vehicles; subsidized freight rates; property taxes related to type of activity; subsidies for mass transit operations; reduced fares on mass transit; permission of higher-density development in exchange for meeting site and building design specifications (density bonuses); congestion tolls; airport noise taxes.

4. Applied to residuals modification, for example: federal and state grants for construction and operation/maintenance of municipal sewage treatment facilities; municipal or state bonds to finance installation of residuals modification equipment in private operations; fast depreciation/tax write-offs on costs of installation of residuals modification equipment in private operations; sewer and landfill user charges; industrial sewer surcharges; reduced taxes for installation of soil erosion reduction measures.

5. Direct public investment in other than residuals modification facilities, for example: open space, highways, mass transit, public buildings, bikeways.

Table 1-2 (Continued)

Administrative--by order within governmental or private agencies

1. Applied directly to residuals, for example; separate various types of paper residuals in offices; collect all used lubricating oil from vehicle fleet for reprocessing; collect all used tires for reprocessing.

2. Applied to products used, for example: specify that only compact automobiles or those which achieve some minimum distance/fuel standard be purchased; specify that only unbleached paper towels be purchased; specify that all writing/printing paper have no more than 60-65 brightness; specify minimum number and shape of containers; specify lighting, energy limits not to be exceeded in new buildings.

3. Applied to activities, for example: specification of limits on thermostat settings for heating and air conditioning; specification that lighting levels within offices after working hours should be $\leq$ some limit.

Judicial

Court and/or administrative law review and action, or threat thereof, to compel compliance; civil and/or criminal suits.

Educational/Informational

Educational/informational programs to acquaint individuals, groups, employees within a firm or agency, of the implications of their activities with respect to residuals generation and adverse impacts on AEQ, and with alternative behavior patterns which would reduce such impacts, for example: car pooling, riding the bus, litter campaigns; provision of technical information by public agencies and private groups to both residuals generators ("technology transfer") and individual citizens. Exhortation by slogan is an example of the "persuasive" implementation incentive, for example: "a good citizen does not litter": "save water--shower with a friend": ""only you can stop waste." Continued major "polluters" can be publicly identified, for example, Region III's "Dozen Dirtiest Dischargers," and the converse, public commendation provided for exemplary residuals management by private and public agencies; appliance labeling programs; fuel economy rating programs; pesticide labeling; model code and ordinance development.

Source: Modified from Blair T. Bower, Charles N. Ehler, and Allen V. Kneese, "Incentives for Managing the Environment", Environmental Service and Technology, 11, 3, p. 253

building incinerators may result in incinerators being closed down, with a consequent increase in solid residuals for disposal.

Fourth, many of the incentives are not under the jurisdiction of environmental quality management agencies. Probably the clearest examples are tax policies such as depletion allowances, capital gains, severance, accelerated depreciation, and real estate.

Fifth, "administrative" implementation incentives reflect internal responses to external stimuli. That is, a public or private manager reacting to external stimuli may adopt an administrative incentive internally within the scope of his jurisdiction. The stimulus may be economic, as when the costs of disposing of solid residuals increase to a level that recycling used paper merits consideration. It may be a public relations move, with or without external pressure from a public interest group. It may be in response to the possibility of a class-action court suit. It may be to set an example for other governmental agencies.

Finally, although this table was developed for the United States context, it is suggestive of the range of implementation incentives which should be considered in any context.

Each implementation incentive has its strengths and weaknesses, so that the implementation incentive(s) chosen in a particular REQM context must be matched to that situation--specific residuals, specific activity categories, specific physical measures. No one implementation is likely to provide optimal REQM at the local, regional, and national scales.

Institutional arrangement. The institutional arrangement for REQM is defined as the set of one or more institutions which has or can obtain the legal authority to impose implementation incentives on residuals generators and to carry out the collective tasks of REQM. Factors affecting the choice of institutional arrangement include: the spatial dimensions of the REQM problem; available implementation incentives; legal considerations; political considerations; and the cultural values of the society. Consideration of the institutional arrangement is often the weakest link in the development of REQM strategies.

Analysis for Regional REQM

Analysis is an integral component of regional REQM. Because analysis generally involves modelling, the function of analysis for regional REQM is referred to as residuals-environmental quality management modelling, which was the focus of the Rotterdam meeting.

Because the nature of the REQM problem and the analytical resources available vary from region to region, the degree of sophistication of regional REQM modelling should be "tailored" to the situation. This is a critical issue which is generally ignored. The difficulty of selecting the degree of sophistication of analysis to use in a given context is compounded by the "intermedia" aspects of REQM. Recognition has grown in the last decade that liquid, solid, gaseous, and energy residuals are physically, technologically, and economically interrelated and that strategies for reducing the discharge of a residual of one form may increase the generation and discharge of residuals of other forms. However, methods for analyzing, and legislation and institutional arrangements for managing, all residuals simultaneously have lagged behind the recognition of the interrelationships and are still predominantly oriented to a single environmental medium.

Developing an REQM strategy is not a process which occurs in a vacuum. Rather, the procedure usually involves some region for which one or more REQM problems have been perceived or identified or within which questions have been raised about REQM. Given such a region the analysis requires: (1) the development of models of residuals generation and modification activities, termed "activity models;" (2) the development of models of the processes which affect, and are affected by, the residuals after their discharge into the environmental media, termed "environmental models"; (3) the specification of an explicit objective function, which includes AEQ indicators either in the function itself or as constraints; (4) the selection of a method of analysis; and (5) the development and application of criteria for evaluating strategies.

Activity Models

For each major residuals-generating activity or group of activities--manufacturing plant, household, municipal incinerator, farm, transport system--an activity model of greater or lesser sophistication is necessary. Such a model: (1) indicates alternative combinations of factor inputs to produce the given outputs of products and/or services; (2) delineates, for different sets of prices of those factor inputs, the types and quantities of residuals generated per unit of activity--ton of steel produced, kilowatt hour of electric energy generated, bushel of corn produced, barrel of crude petroleum processed, inhabitant in household; and (3) identifies the various physical measures available for reducing the discharge of residuals into the environment and the costs of various degrees of discharge reduction. Depending on the time and resources available and on the relative importance of the different generators for the given REQM context, activity models may be highly aggregated and simple, or highly disaggregated and complex, and may take the form of mathematical models.

Environmental Models

The outputs of the activity models--types and quantities of residuals discharged at specific locations and times--comprise the inputs into the environmental models, along with the relevant hydrologic, geomorphologic, meteorologic, and pedologic variables, such as temperatures, wind velocity, precipitation, soil characteristics, topographic slope, stream channel characteristics, and sunlight. Major types of environmental models are: (1) physical dispersion models such as for suspended particulates, sulphur dioxide (SO_2), and sediment runoff; (2) chemico-physical dispersion models such as for photochemical smog, and nitrate movement and modification in ground water aquifers; and (3) biological systems models such as terrestrial and aquatic ecosystem models. Just as for activity models, the time and resources available and the relative importance of the environmental media--in terms of the relevant AEQ problems--determine the degree of complexity of the environmental models. For example, a water quality model may consist of a set of simple linear transfer coefficients or it may be a multi-compartment aquatic ecosystem model. Environmental models transform the time and spatial pattern of residuals discharges into the environment into the resulting time and spatial pattern of AEQ, as measured by whatever indicators are of interest in the particular case.

Objective Function and Method of Analysis

An objective function is a statement of the criterion or the criteria for which the best solution is desired. Constraints are additional factors which must be taken into account in the solution of the problem, factors which limit the range of permissible solutions. Criteria can be reflected in the specification of constraints as well as in the specification of the objective function. Constraints are also part of the decision-making criteria. Specification of an explicit objective function and the accompanying constraints serves to translate the goals and/or specific objectives of a study into a mathematical or quantitative description of the REQM problem.

If damage functions are available which translate AEQ into monetary damages to receptors, such as costs associated with impacts on human health, value of yield loss for agricultural crops, costs of cleaning buildings, then the objective function can be expressed as, maximize the present value of net benefits--the present value of the time stream of damages reduced minus the present value of the time stream of REQM costs, over some specified time horizon with the relevant social rate of discount.

Developing damage function involves the determination of the relationship between short-run and/or long-run exposure to the concentration of a residual and physical or physiological effects, and then the determination of the relationship between those effects and monetary costs. Such determinations are difficult for at least three reasons. One, most damage functions involve multiple variables in addition to the residuals concentration-duration variable. For example, the effect on alfalfa of a given concentration of any one of several gaseous residuals is higher at high humidities than at low humidities. The impacts on human health of SO_2, particulates and probably other gaseous residuals depend on other variables related to the receptor, such as the general physical condition, age, nutrition and smoking habits. Two, often several residuals of the same form or in a single environmental medium are simultaneously involved, such as SO_2, particulates, and nitrogen oxides, so that there may be additive or mitigating effects. Three, the same residual, such as lead, can impact the individual through ingestion into lungs in gaseous form, through liquid intake, and in foods. Determining the relative impacts of each on the "total body burden" is difficult.

Generally, all the necessary damage functions are not available to enable utilization of a net benefit objective function in regional REQM modelling. Recourse must then be made to some form of cost minimization objective function, such as minimize the present value of costs to meet a set of AEQ standards and/or limits on discharges, with or without a capital or capital + O&M constraint, and/or with other constraints.

Once the objective function has been specified, two basic methods of analysis are available. One involves the use of mathematical programming, such as linear programming. Given: (1) a level of output for each activity category; (2) the costs of different degrees of individual and collective residuals modification activities; and (3) the costs of different degrees of direct modification of AEQ, such programming specifies the minimum cost set of physical measures to achieve the specified AEQ standards and/or to meet specified upper limits on residuals discharges.

The other method can be characterized as a manual "search" method. It is illustrated in figure 1-3. Given the specified level of output for each activity category, one or more discrete options for reducing residuals discharges and the associated costs are delineated for each activity category and for collective residuals handling and modification activities. As with the programming method, costs of different degrees of direct modification of AEQ are also delineated. Various combinations of these options are then specified, and the effects on residuals discharges, on AEQ, and on REQM costs

Figure 1-3. Flow Diagram of Search Method of Analysis

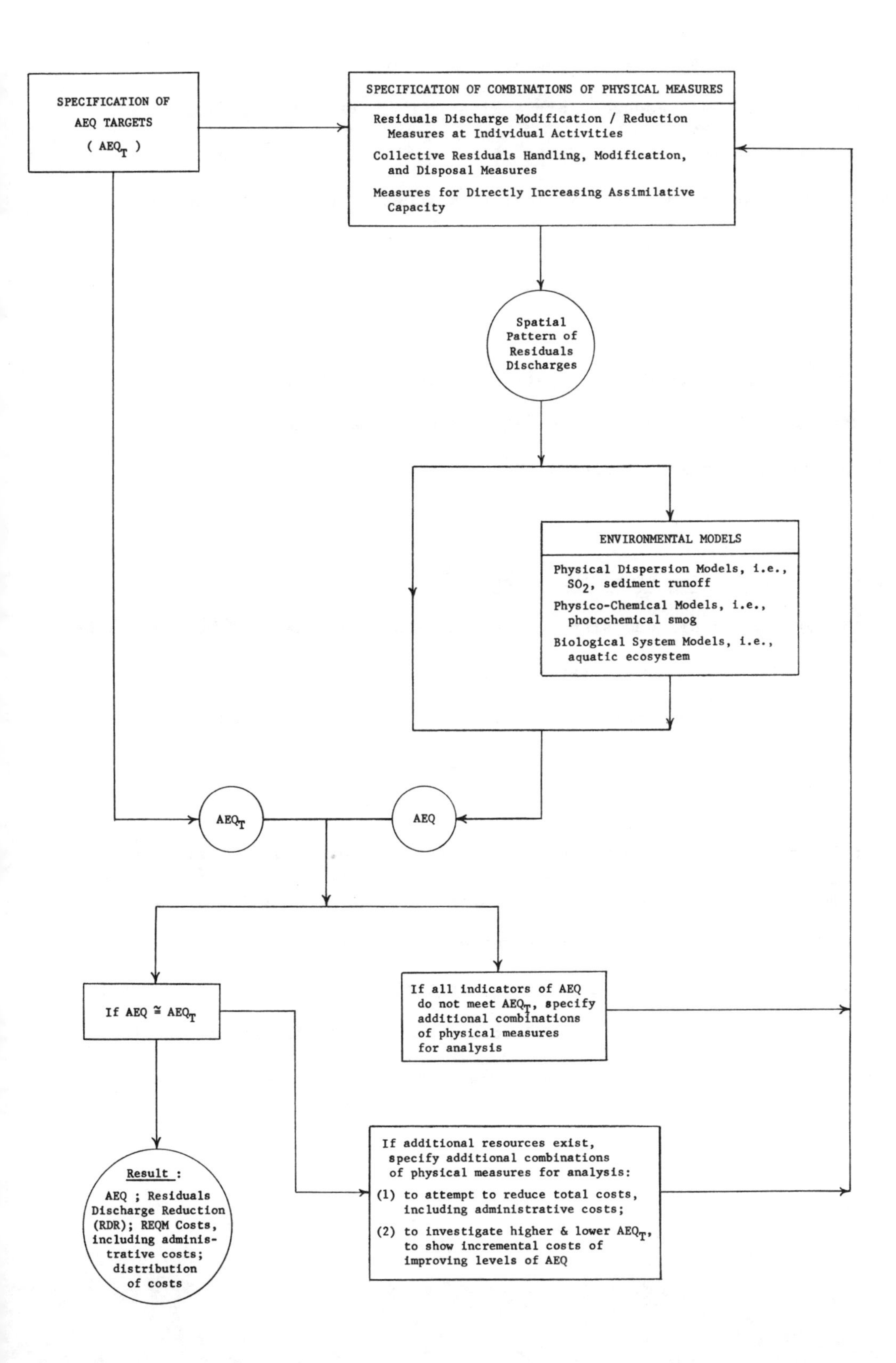

are tabulated. The analysis of possible combinations is continued as long as computational resources are available. Rarely are sufficient resources available to make a search of all possible combinations. However, there are formal mathematical techniques which can be used to sample the response surface.[3]

Both methods of analysis enable investigation of the effects of the following variables on residuals generation and hence on AEQ in the region and on REQM costs: alternative spatial patterns of activities; alternative final demands; alternative transportation systems; alternative collective residuals handling/disposal facilities; and alternative AEQ standards and discharge constraints.

Criteria for Evaluating REQM Strategies

In REQM, as in all decision-making contexts, criteria must be established by which to choose an REQM strategy. These critera represent factors which decision makers consider relevant in evaluating strategies. Not only must the criteria be specified but relative weights must be attached to them.

Although real resource costs of an REQM strategy represent a major factor in choosing a strategy, these costs are not the only criterion. Decision makers in all societies use multiple criteria in making decisions, with the criteria and their relative weights (importance) being made more or less explicit. Table 1-3 lists a set of criteria for evaluating REQM strategies. These criteria may be applied to each residual/activity category/physical measure/implementation incentive combination, or to each REQM strategy as a whole.

Table 1-3. Criteria for Evaluating REQM Strategies

1) Physical Effects	2) Economic Effects
3) Flexibility in Administration	4) Timing Considerations
5) Political Considerations	6) Intermedia and Resource Use Effects
7) Accuracy of Estimates	

1. Physical effects relate to the degree that the physical measure will: (a) reduce the discharge of a residual from a specific source category; (b) reduce the total discharge of the residual in the REQM region; and (c) change the relevant indicator of AEQ; these changes in discharge and AEQ can, in turn, result in (d) other physical effects such as decreased human mortality, decreased human morbidity, decreased deterioration of materials, and increased fish biomass.

2. Economic effects include: (a) direct benefits--the monetary value of the changes in physical effects, such as reduced medical costs, reduced costs of cleaning and maintaining structures, and increased value of fish catch; (b) the direct costs--capital and O&M--to the residuals discharger, individual or collective, of implementing the physical measures for reducing discharges, and to the REQM agency for implementing physical methods directly affecting assimilative capacity; (c) public and private administrative costs for accounting and reporting, monitoring, analysis of samples, supervision of operating personnel; and (d) indirect benefits and costs as reflected in employment effects, changes in income taxes, changes in property taxes, increased costs of user goods and dislocation of people. Although administrative costs are direct costs, they are

separately identified because they are too often ignored.

A very important consideration with respect to economic as well as physical effects is their distribution. Who benefits from improved AEQ and who pays in what forms for that improvement? Distributional effects should be determined in relation to: (a) political jurisdictions and socioeconomic groups of the population within the REQM region; and (b) the division between direct costs incurred <u>within</u> the area and incurred <u>external</u> to the area. In many contexts another important consideration with respect to costs is the extent of foreign exchange required by the given strategy.

3. <u>Flexibility in administration</u> refers to the administrative ease with which an implementation incentive related to a particular physical measure-activity category combination may be imposed or removed, and to the degree to which it remains effective for that combination under changing conditions. Changing conditions include: (a) seasonal variations in both residuals generation and assimilative capacity; (b) changes in factor prices, such as fuel, energy, water; (c) changes in technology over time; (d) new information, such as with respect to the behavior of the natural systems involved and the behavior of residuals generators in responding to implementation incentives and to changing factor prices; and (e) new social goals and priorities.

Two important attributes of flexibility relate to whether an implementation incentive can be applied to activities: (a) continuously or noncontinuously; and (b) selectively or uniformly. Continuous/noncontinuous refers to whether the physical measure/implementation incentive combination can only be applied continuously or can be applied intermittently as needed.

Selective/uniform refers to whether or not the physical measure/implementation incentive combination can be applied to selected activities--either within a category or among categories--or can only be applied to all activities generating the residual of interest. In some institutional contexts it may be administratively simpler and politically easier to implement the adoption of a physical measure by all activities than to impose it selectively, because uniformity has a "ring" of equal treatment. Total REQM costs are almost always higher with uniformity.

4. Timing considerations relate to the fact that physical measure/implementation incentive combinations vary with respect to both the time required to install the physical measure and place it in operation and the time required after it is in operation before the effect on AEQ occurs. Timing may be a particularly important consideration where there are adverse AEQ conditions which need to be ameliorated as soon as possible. Timing is affected also by legal considerations. If new legislation must be enacted, implementation may take longer than if legal authority already exists.

5. The political considerations criterion has five components. The first refers to the perceived urgency of the particular REQM problem in relation to other REQM problems, e.g., improved air quality vis-a-vis improved water quality, or in relation to other locations of the same REQM problem. The second refers to the perceived urgency of REQM problems in relation to other societal problems in the region. The third is the impact on intergovernmental relations, e.g., federal-state, state-local, inter-local, in terms of the strategy's effect on the normal way of carrying out the government's business. The fourth component is public acceptance.

A physical measure/implementation incentive/institutional arrangement combination which is new and/or unexplained to the public may inhibit acceptance. Presumably public acceptance is most easily gained through the involvement of the public from the initiation of, and throughout, the planning process. The fifth component relates to the degree of difficulty in obtaining legal authority for the institutional arrangement to implement the strategy. Does adequate authority to implement the strategy exist; could existing legislation be changed to enable implementation; would entirely new legislation have to be passed or executive decrees promulgated?

6. Intermedia and Resource Use Effects should be tabulated in physical terms, as well as being incorporated in economic costs. Intermedia effects refer to the quantities of additional residuals generated and discharged into any of the environmental media as a result of application of the strategy. The three resource use effects most often of interest are: net energy required; net land required; and net consumptive use of water. An REQM strategy may be energy intensive; or it may actually reduce total energy use in the region. This is an important factor where a significant proportion of the basic energy supply, or of a particular type of fuel--such as petroleum--must be imported. The use of ponds, lagoons, and/or spray irrigation to reduce liquid residuals discharges may increase the net consumptive use of water in a region. The importance of the land required for an REQM strategy in a densely urbanized area, for example for disposal of mixed solid residuals and sludge, may be inadequately reflected in land cost calculations.

7. Accuracy of the estimates of the costs of the REQM strategy and of

the impacts on AEQ which the strategy is predicted to have may affect the choice of strategy. A strategy which has large estimated positive effects on AEQ and/or low costs, but for which there is large uncertainty in the estimated costs or effects or both, may not be preferred to one which has substantially less impact on AEQ and higher costs but for which the probability of achieving those effects is high.

After evaluating each strategy according to each of the indicated criteria, the final step is to combine the ratings for the individual criteria. This process involves assigning relative weights to the individual criteria, an activity which is the responsibility of the decision makers, not of the analysts.

Concluding Comment

The preceding discussion has: defined residual and some basic concepts related thereto; described the elements of REQM and REQM strategies; and described the procedure of analysis for regional REQM. It is obvious that, whatever the structure of a society, decisions will be made with respect to levels of AEQ desired and choices will be made of strategies to achieve them. Regional REQM modelling generates information necessary for those decisions and choices. As an analytical procedure, such modelling is apolitical. What differs among societies and among contexts within a society are the goals, social values, constraints, and the analytical resources available to carry out the procedure. This apolitical nature of the analytical procedure is exemplified by the case studies presented to, and was reflected in the discussions at, the Rotterdam meeting. These are reported in the following chapters.

Footnotes

[1] Various examples of the interralationships among these three factors exist. DDT and other chlorinated hydrocarbons were under suspicion some time before instruments were developed which were sensitive enough to detect the presence of chlorinated hydrocarbons in the low concentrations in which these materials are typically found in the environment. Nitrates in ground water represent the reverse situation. Although methods of measuring nitrates in water were long available, it was not until their effects on babies and on accelerating eutrophication in some cases became known, that nitrate concentration was utilized as an indicator of AEQ.

[2] The result of such measures may be to increase ground level concentrations some distance away outside the region.

[3] See Maynard M. Hufschmidt, "Analysis by Simulation: Examination of Response Surface", in A.A. Maass, et al, Design of Water-Resource Systems (Cambridge, Mass: Harvard University Press, 1962).

Chapter 2

ISSUES SURROUNDING REGIONAL RESIDUALS-ENVIRONMENTAL QUALITY MANAGEMENT MODELLING

Allen V. Kneese and Blair T. Bower

"When we mean to build,
We first survey the plot, then draw the model;
And when we see the figure of the house,
Then we must rate the cost of the erection;
Which if we find outweighs ability,
What do we then but draw anew the model
In fewer offices, or at last desist
To build at all? Much more, in this great work,
Which is almost to pluck a kingdom down
And set another up, should we survey
The plot of situation and the model,
Consent upon a sure foundation,
Question surveyors, know our own estate,
How able such a work to undergo,
To weigh against his opposite; or else
We fortify in paper and in figures,
Using the names of men instead of men:
Like one that draws the model of a house
Beyond his power to build it; who, half through,
Gives o'er and leaves his part-created cost
A naked subject to the weeping clouds
And waste for churlish winter' tyranny."

William Shakespeare
Henry IV, Part II, Act I, Scene III

"In the space of one hundred and seventy-six years the Lower Mississippi has shortened itself 242 miles. That is an average of a trifle over one mile and a third per year. Therefore, any calm person who is not blind or idiotic, can see that in the old Oolitic Silurian Period, just a million years ago next November, the Lower Mississippi River was upward of one million three hundred thousand miles long. By the same token any person can see that seven hundred and forty-two years from now the Lower Mississippi will be only a mile and three quarters long. There is something fascinating about science. One gets such wholesale returns of conjecture out of such a trifling investment of fact."

Mark Twain

"Economic efficiency is not a cause which brings men cheering into the streets."

Richard S. Howe

Introduction

This chapter is not meant to provide conclusions about the issues surrounding regional REQM modelling. Its purpose is to raise a number of questions for discussion, beginning with some general issues, then moving to more specific issues. The case studies which follow shed light on how some of these issues were tackled in specific modelling contexts. The meeting report summarizes the results of the discussions during the Rotterdam meeting.

The general issue is, of course, why one would wish to approach REQM problems on a regional basis at all. A quick and frequent answer, but not a wholly convincing one, is that the various "problem sheds" are regional in extent. By this is meant that they do not conform to the usual national, state, or local political boundaries, but rather are delimited by the natural systems occurring in watersheds, ground water basins, coastal areas, and atmospheric regions.

It is not so clear, however, that the issue is so simple, because there are factors relevant to the decision making other than the extent of the natural systems, even where these systems can be defined with a relatively high degree of clarity. There may be at least three different economic-ecological type regions which are pertinent to this question. First, there is the source and control region where discharges of residuals are made and control over discharges and over some environmental impacts might be exercised. This type of region may be centered in a metropolitan area-industrial complex, an agricultural area, a mining area. Second, there is the "effects" region which is better bounded by a natural environmental system of some type--a watershed, a ground water basin, an atmospheric region--although the long-distance transport of some residuals may make the natural environmental system involved too large to "handle." Finally, there is the broader economic region which may be impacted by activities in the source and control region which affect environmental systems. For example, the Rhine River is of concern to the countries in its watershed, but, because of its great and highly valued scenic attributes, it is also of interest to some parts of the population throughout the entire world.

Futhermore, some analysts and practitioners would give primacy to the political concept of a region. That is, they would argue that political realities dictate that REQM problems be approached by existing governmental units at the national, state or provincial, and local levels. Whether these political boundaries correspond to some definition of a natural or economic region is beside the point, they argue, because acceptable institutional arrangements at the regional level can rarely be achieved in any society. The type of region for which REQM studies have so far been done, has tended to be defined by at least a loose conception of some sort of natural environmental system. In some cases the problem of political jurisdiction has been regarded as part of the research question, i.e., an effort was made to use engineering, economic, and ecological analysis combined with political analysis to address questions of optimum jurisdiction and the appropriate political structuring of decision making. The Rotterdam meeting focused on the issues surrounding studies embodying technological, economic, ecological, and institutional aspects in regions corresponding at least roughly to the boundaries of natural systems, but with explicit attention being given to the question of how management boundaries are best defined.

A second general issue is whether or not to analyze simultaneously all forms of residuals--waterborne, airborne, and solid. The traditional approach was, and still often is, to analyze each environmental medium separately, ignoring intermedia linkages and the implications of inter-form transformations. There are at least three types of intermedia linkages which suggest the importance of integrated analysis and management. One is that a change in one or more of the production process variables of an activity may change two, or all three forms of residuals, as when the brightness specifications on paper products or the specifications on fuel inputs are changed. A second type of linkage is that which occurs when the modification of one residual also results in the modification of another, as when in-plant recirculation of heated water--in order to reduce thermal discharges--results in a reduction in fuels use, thereby reducing both gaseous and solid residuals generation. A third type of linkage is the generation of a secondary residual in the process of modifying the originally generated residual, as in the generation of sludge in the removal of BOD, the generation of fly ash in the removal of particulates from a gaseous stream. Of course the sequence may--and often does--proceed further, as in the generation of particulates in the incineration of sludge, and the generation of a solid residual in removing these particulates from the incinerator stack. (In addition it is likely that additional gaseous residuals are generated in fuel combustion to generate the energy for the residuals modification processes.)

A third general issue in regional REQM modelling involves the evaluation of alternative strategies for improving AEQ, where a strategy involves: (1) a set of physical measures to reduce discharges and/or to improve AEQ directly;

(2) the related set of implementation incentives to induce the adoption of the indicated physical measures; and (3) the institutional arrangement which defines the governmental responsibilities for imposing the implementation incentives and for carrying out collective REQM activities. One of the criteria for defining the utility of regional REQM modelling is the extent to which "light" can be shed on the costs and consequences of alternative REQM strategies.

A fourth general issue involves the "payoffs" to increasing the refinement, sophistication, complexity of regional REQM models. The real world is stochastic, dynamic over time and space, nonlinear. How far toward reproducing these characteristics does regional REQM modelling have to go to be useful in decision making? How do the costs of additional refinement compare with the values of the additional information produced? Or is this issue irrelevant because the model builders are only talking to themselves and each other, while the decision makers ignore simple and complex models alike?

Issue A: Potential Use and Users

While the existing examples of regional REQM models have usually been constructed with the help and encouragement of staff members of some existing unit or units of government--at least with respect to provision of data--they have in general reached beyond the missions of existing agencies or beyond the range of options perceived as feasible by those agencies. Because there is, therefore, not necessarily a set of "ready-made" users for the outputs of these modelling efforts, a question is raised as to who will use the results of such regional REQM models, and perhaps the models themselves. Among the various parties potentially using them do different ones need different types of output and which types of users need which types of output? It may be, for example, that one set of users will be other researchers and the technical staffs of management agencies. These presumably will need access to the full mathematical, empirical, and computational details of the regional model. On the other hand, political decision makers and higher-level administrators may be able to use only highly aggregated summary information which sheds light on major alternative management strategies or the broad environmental, economic, and political implications of particular decisions. It is important then to give close consideration to the question of how outputs should be organized for, and presented to, the different users. This may have substantial impact on the delineation of the model, for example, if the given user is particularly interested in the question of who pays the costs of improving AEQ and who gains from that improvement. Different users may well want different questions answered.

A related question is what makes a regional REQM model "believable" to potential users of the outputs. If the results are not believed, then of course they will not be used, except perhaps cynically for strategic purposes. This raises the very important question of what constitutes a reasonable verification of regional REQM models. Because they are intended to estimate what will happen if new policies or new approaches are adopted by the governmental agencies responsible, the outputs can never be fully verified short of actual real-world experiments. These would be very time consuming and extremely expensive, and perhaps therefore, unacceptable. Because most models to date are of a steady-state variety, there is also a question of exactly how one matches results to the dynamics of a real-world situation. Are there lesser kinds of verifications, based perhaps on historical simulation or other approaches, which will suffice to make users believe that the model results have a reasonable fidelity to reality?

Issue B: Elaboration of Economic Models

Several major sets of questions fall under this heading. The first is the question of general equilibrium models versus partial equilibrium models. The second set surrounds the connections between a particular region and regions to which it is economically linked, i.e., the question of interregional models. The third involves the question of how detailed models of industrial and other residuals generating and discharging activities can, and should, be. A fourth set might involve the interrelations among micro, regional, and macro models.

General Versus Partial Equilibrium. Many ecologists state that everything is connected to everything else--a statement which contains some truth, but which is almost totally unhelpful in an operational sense. Even economists realize that there are numerous nonmarket linkages in economic systems, which might make it useful to undertake general equilibrium types of approaches to residuals management. To the extent that one regards an input/output (I/O) model as a bonafide general equilibrium model, some studies have been made which do use a general equilibrium approach within a given region. In general to make this possible it has been necessary to sacrifice detail and the ability to generate realistic and rather accurate policy analysis. On the other hand, the partial equilibrium approaches which have been characteristic of some of the major regional REQM studies do make cuts which may or may not be justified. For example, they do not consider the effects which an increase in output in one industry may have on residuals generation and residuals discharge in another industry due to the increased demand for intermediate input. Accordingly, the question arises of the comparative benefits and costs of the two approaches. Are there operational techniques for attaching, for example, I/O models to partial equilibrium models in such ways as to obtain some of the benefits of both approaches? How can environmental models of different types be incorporated within each approach? It should be noted that I/O models in particular seem to have had great difficulty in making reasonable linkages to environmental models. One can envisage a continuum of tradeoffs between the complexity, i.e., degree of disaggregation of the I/O model, and the complexity of the environmental models combined with it. The optimal combination presumably is that which yields the greatest "worth" of the output data to the user, in relation to the costs of analysis, i.e., obtaining information. The optimal combination is of course a function of the questions the user wants answered.

Interregional Models. In the same way that inter-industry relationships exist and hence may present problems for partial equilibrium analysis within a region, such linkages connect a particular region with other regions in various degrees of tightness, which may be a problem for interpreting the results of both partial and general equilibrium analyses of a given region. This is especially so when factors affecting industrial and agricultural location are explicitly taken into account. The question arises as to whether interregional models have been or can be developed on a detailed applied level suitable for regional REQM analysis. How much is to be gained by developing such models? On what sort of policy issues would they shed light that regional REQM models cannot by themselves? Can and should such models be developed in a way that would aid location decisions? What are the problems of combining interregional "industry" models and regional REQM models? Again, one might envisage tradeoffs between these two components of an analysis, with the combination chosen a function of the questions being asked.

Models of Industrial and Other Activities. Some form of model of production and use activities which generate and discharge residuals is a necessary part of any regional REQM modelling. In practice such models have ranged from fixed coefficient types, as in I/O models, to fairly elaborate optimization-type submodels of major industrial and other activities. The latter have taken the form of linear programming and simulation models, and a relevant question is what are the advantages and disadvantages of each of these approaches. This includes such matters as how cost functions involving interacting variables can be handled and how non-convexities can be handled. Another related question is which type of formulation of the activity models makes them more readily usable in the larger regional models. A major problem in practice has been how to simplify the detailed models of economic activities in such a way that they could be incorporated in regional REQM models without creating impossibly large computational difficulties. In view of limits on computational capacity and of the higher cost of running large models, a critically important question is how this best can be done.

To date, most of the activity modelling has concentrated on industrial activities. Little work has been done on the so-called nonpoint source activities, such as agricultural operations, forestry operations, and urban storm runoff, and on transportation activities. In some regions these activities are the primary source of residuals discharges to the environment, and hence of deterioration in AEQ. While there are no conceptual problems in developing models of these activities, empirical work is badly needed. In addition, the municipal sector--itself a mix of activities--merits more attention, particularly in view of the interest, on the part of at least a few governmental agencies directly responsible for REQM, in finding measures to modify municipal "demands" for environmental services.

A final issue with respect to the elaboration of economic models involves the utility of integrating macro, regional, and micro level analyses. Some attempts have been made in this direction, and some are underway, with the primary emphasis having been, and being, on the macro segment. The "numbers" being ground out are being used in policy deliberations, even though it is not at all clear how good the numbers are. Full-fledged integration of a detailed I/O national model with regional disaggregation--implying some interregional model--with linear programming models of activities within regions with nonlinear environmental models, "boggles" the mind (and would delight computer programmers and IBM stockholders).

Issue C: Elaboration of Environmental Models

The classic type of environmental model which has been used in regional REQM modelling for both water and air is a linear (transfer function) model of the relevant natural system. Such models are based on underlying (primitive) nonlinear equations, which assume linear form when solved for steady-state conditions. It is fair to say that the use of such models has been very illuminating in terms of REQM policy alternatives and has served a useful purpose in planning and implementation activities in particular regions. They are, however, radical simplifications of the behavior of the actual systems, in that they assume steady-state conditions and do not explicitly include a number of relevant variables, and their outputs--being in physical or chemical terms--may be difficult to relate to the actual values being adversely affected by residuals discharges. Accordingly, some efforts have been made to develop more elaborate and, in principle, more realistic models. This effort is perhaps most well developed in the aquatic area.

Aquatic models. Conventionally, Streeter-Phelps type models, elaborated to incorporate multiple dischargers and multiple receptors, have been used in regional water quality management studies. More recently aquatic ecosystem models have been developed which, in addition to parameters such as dissolved oxygen, estimate impacts on living species such as fish and algae. The latter-type models are much more costly in terms of data requirements and computational time than the former. This raises the question of what are the benefits of each type of model in terms of range of output, accuracy of output, useability of output, and reliability of solution. Each of these considerations must be measured in relation to data needs, costs of data collection, and computation costs. The basic principle is that a model (environmental or activity) should be no more complicated than is necessary to provide the needed information with acceptable accuracy. Ergo, under what conditions are simple linear models sufficient; when should the more complex aquatic ecosystem models be used; what degree of complexity is justified?

Because all models so far used rely on steady-state solutions, an important question is how one best selects the steady-state conditions and how meaningful a steady-state is in relation to any ecological system. In some situations the most important question for real-world decision-makers may be the extent to which, and at what costs in terms of external inputs, an aquatic ecosystem can recover from shock loads. In other cases the relevant question is, to what extent and at what cost can any sort of biological system be reestablished in an essentially anaerobic stream.

On a broader level there is the question of the relative importance of three-dimensional models versus two-dimensional models versus one-dimensional models (where dimension refers to space). This question is particularly important for bays, some estuaries, and many lakes, where significantly different conditions exist in all three directions.

Air Quality Models. Linear dispersion type models have been used extensively in the definition and analysis of air quality management. They are limited to non-interacting--or approximately non-interacting--residuals such as SO_2, suspended particulates, and carbon monoxide. An important question is how useful have these linear atmospheric models actually been in terms of accuracy of output and solution reliability, particularly in urban areas. Are efforts to increase the accuracy of estimating concentrations over space worthwhile in terms of the outputs they yield and the additional costs of obtaining them? Do they adequately take into account "washout" phenomena?

Some very important atmospheric quality degradation phenomena cannot be successfully modelled at all by the use of linear dispersion techniques. Some efforts have been made to model effects of residuals involving chemical processes in the atmosphere. What are the prospects of such models, such as for photochemical smog, in the context of regional REQM modelling and decision making for REQM? Another residual which dissipates in the atmosphere is heat. This residual is becoming increasingly important in major urban areas, in terms of its effect on the climate in such areas. Pressures to prevent or substantially reduce thermal discharges to watercourses are resulting in large quantities of heat discharge to the atmosphere. For example, one proposal has been made to install cooling towers for 10,000-15,000 MWE nuclear power plants for an urban region in a topograhically enclosed basin. It is not clear how well thermal discharges to the atmosphere can be modelled.

Terrestrial Models. Virtually all of the work to date on environmental models in the context of regional REQM modelling has been on models of aquatic and atmospheric systems. Yet even in urban areas there can be substantial contributions to residuals discharges from overland flow. In areas where agricultural or forestry operations predominate, terrestrial models are clearly essential. What is the state of development of such models? How well can they predict the transport, transformation, and final disposition of pesticides, fertilizers, suspended sediment?

Issue D: The Spatial Dimension

The matter of how space is to be handled arises in several respects. In the opening paragraphs references have already been made to the boundary, or optimal jurisdiction, set of problems. Within the boundary selected there is the question of how sites are to be located. That is, what activities are to be located where, and can and should such decisions be made internal to (a decision variable in) the regional REQM model? Finally, in view of the numbers of discharge locations found in any moderately complex region, the question of whether one should treat each discharge point discretely, or should aggregate them in various ways, is extremely important.

The Boundary Problem. The question is, how should the boundaries of the region be determined and how do these boundaries relate to the jurisdictional boundaries of existing governmental agencies? Can regional REQM modelling shed useful light on how existing agencies might be restructured to reflect better both the preferences of people affected and the technologies available for REQM?

No boundary can possibly be defined which internalizes all effects of decisions within it. Residuals will "spill out" and people not represented in the region's political processes will "spill in." (Residuals may also "spill in," as in the transport of gaseous residuals.) How are these boundary problems best handled? Can regional REQM models say anything useful about the appropriate hierarchical relationships among different levels of government with respect to REQM problems? This last question is particularly important in view of: (a) the wide range in degree of centralization of decision making with respect to REQM which exists; and (b) the range in number of governmental levels involved in REQM in any given region.

Site Location. Regional REQM models have generally taken the sites of activities as given, or else permitted external manipulation of sites with consequent testing of some of the implications of alternative spatial distributions of activities. Because the location of activities can frequently have an important effect on AEQ, and hence on the costs of REQM, a question must be addressed as to what are the tradeoffs, difficulty versus utility, involved in including site location variables as explicit decision variables in regional REQM modelling. Can simulation analysis be used either as a complement, supplement, or substitute for optimization analysis in the examination of the site location problem? What are the consequences for regional REQM of substituting height for area in the location of activities? Or is this a level of detail which can be adequately handled by changes in coefficients, based on some "side" calculations and/or assumptions?

Discrete Points Versus Aggregation. Because, as already noted, the specific points of discharge in a region are often very numerous (which creates complexities for modelling, data collection, and analysis), several important questions arise. A straightforward one is, how important is it to maintain discrete point sources of discharges as inputs to environmental models? A closely related question is, what sizes should grids, water course reaches, and ground water basin volumes defined for modelling purposes be? Similarly, how should areas of nonpoint sources be delineated? If aggregation is to be undertaken is there any evidence from previous efforts as to what techniques are most promising? How much information is lost? How much is gained in terms of reduced computation, data collection, and analysis time? What degrees of aggregation significantly affect what types of policy and management analyses? More generally, what are the relative costs and benefits of different types and degrees of aggregation?

Issue E: The Time Dimension

Regional REQM models have tended to be of a static, steady-state nature, or in some cases of a comparative static nature. There are also efforts in early stages of development to consider some stochastic aspects. Both the dynamic and stochastic aspects of regional REQM modelling and of regional REQM are important.

In general, the dynamic aspects of changes in population, economic activities, and ecological systems have not been handled or have been treated only very inadequately. A number of questions assert themselves in this connection. How important and how difficult is it to include changes in the magnitude of regional activities over time, with respect to problems of capacity determination and expansion; to include changes in the spatial distribution of activities in the region over time? How can changes over time in technologies of production processes, in final demand, in characteristics of raw material be handled in regional REQM modelling? (Macro and regional I/O models cope with the problem at least partially by changing technical coefficients and using dynamic demand functions.) How can the accumulation of residuals in the environment and the time dependent and perhaps difficult to reverse effects of such accumulation be incorporated into regional REQM models?

Nearly all variables in the regional REQM problem are stochastic in nature. These include both the human-influenced ones and those that are the result of the behavior of natural systems. Thus, for a given set of activities distributed over space in a region--that is, production processes, plant capacities, inventory of dwelling units, etc., fixed--there are normal variations which result in a variable pattern of residuals discharges. Similarly, the natural systems vary diurnally, seasonally, from year to year. The question then is, how important and how difficult is it to take explicitly into account in modelling the probabalistic elements of fluctuations in demand, production, residuals generation, residuals modification efficiency, residuals discharge, and the fluctuations in nature? How can investment and operating problems which vary with seasonal differences in the environment be handled?

Issue F: Model Structure and Mathematical Techniques

The selection of a model structure and techniques for solution has many ramifications, some of which have already been implied by issues raised earlier. A central decision in this respect, shaping in one way or another all subsequent modelling, data collection, data arrangement, and analysis activities, is whether optimization or simulation should be chosen, or some combination of the two. Optimization is more confining in terms of the model structure it imposes, but it does involve a known maximum-seeking technique which can be extremely helpful in coping with the vast complexities of regional REQM modelling. Simulation, on the other hand, is much more flexible in its ability to handle complicated relationships but may present major difficulties in implementation and interpretation. What are the specific advantages and disadvantages of each and to what particular types of situations may each be appropriate?

If the optimization model route is chosen, there still exists the question of whether such a model should be entirely linear or incorporate nonlinear elements. While the possibility of including certain types of nonlinear elements has been clearly demonstrated, there are nevertheless remaining problems. One of course is that the amount of computation is greatly increased if nonlinear elements are introduced. Is it worth it? Linear optimization models, with the constraint of being linear, do tend to find the global optimum. In large complex models nonlinear elements may cast doubt upon whether a local or global optimum has been found. Is this added uncertainty worthwhile? Are there practical means of testing which type of optimum a model with nonlinear elements has reached?

It is quite possible that some technologies for handling residuals will have concave cost functions. This may be true, for example, of collective residuals handling and modification facilities and possibly facilities for modifying the environment itself, e.g., instream reaeration. What are the possibilities, advantages, and disadvantages of different model structures and mathematical techniques in handling such situations? There also can be problems with discontinuous cost functions relating to residuals modification at individual activities. How can these best be handled?

A similarly large question relates to whether the model should be able to handle all types of residuals simultaneously or whether taking residuals one at a time suffices. This question involves the matter of how strong are the linkages and hence how important are the tradeoffs among types and forms of residuals and consequently among the environmental media to which they can be discharged. What light is shed on this question by the modelling work already done?

Regional REQM models often take the general form of economic benefit-cost analysis, granted, of an unusually complex type. Full implementation of this mode of analysis requires the estimation of damage functions. What is the state of the art with respect to such functions, both from a conceptual standpoint and from an empirical standpoint? What are the arguments for and against their explicit inclusion in regional REQM models? If they are not included, what other approaches can be used to demonstrate the impacts of alternative strategies and the distributions of costs and gains?

Finally, computation inevitably arises as a sticky problem in large models. How is this problem best handled? Are decomposition techniques, such as hierarchical programs, the best approach? Is simplification of the overall modelling approach indicated? Are more efficient computation algorithms likely to be helpful? Are some existing computational techniques and computer systems clearly superior to others now in use? Clearly, the overall question involves the tradeoff between comprehensiveness and computational cost.

Summation

This is the section for broad perspective-type questions. Where do we stand? Where do we go from here? Is the tendency toward larger and more complex models a desirable one? Are present models adequate and should we focus our attention now on empirical problems and applications? Is it demonstrated that formal models of regional REQM are superior to engineering and administrative judgments concerning what is the "best" REQM strategy? Under what conditions should what type of regional REQM model be used to answer what questions? For example, if the objective of the analysis is to assess roughly the relative importance of different residuals generators and their impacts on AEQ, the regional REQM model would have to be less detailed than if the objective were to select a strategy for actual application in a given region. Are there criteria which would help in the model selection process, i.e., relating to the interrelationships among purpose of the analysis, degree of detail required for the purpose, and analytical resources available? Can tradeoff functions among comprehensiveness of models, worth of outputs, data, and computation costs be defined? Of what help might regional REQM models be in the day-to-day operation of managing AEQ in a region, a problem totally ignored in the previous sections?

The above enumeration of issues provided more questions than could be answered in the four days of discussions at Rotterdam. However, it is hoped that the discussions stimulated by the questions did result in an incremental step forward toward useful analysis for regional REQM decision making.

Chapter 3

THE LOWER DELAWARE VALLEY INTEGRATED RESIDUALS MANAGEMENT MODEL: A SUMMARY

Walter O. Spofford, Jr., Clifford S. Russell,
and Robert A. Kelly

Introduction

This paper describes the essential elements and application of a regional integrated residuals management model developed at Resources for the Future by an interdisciplinary team representing the fields of economics, ecology, engineering, and political science. The illustrative application to the Lower Delaware Valley represents the final phase--an application in the real world--of a research effort at Resources for the Future which has concentrated on the development of regional residuals management models to aid government in establishing public policy with respect to regional environmental quality--air, water, and land. A number of publications written by members of the modelling team describe in some detail the various stages in the development of the regional residuals management model.[1]

The summary of the Lower Delaware Valley residuals management study is organized as follows: the objectives of the study; a résumé of the region and its residuals environmental quality problems; the regional residuals management model; the modelling studies conducted; the results achieved; and conclusions.

Objectives of the Study

The application of the regional residuals management framework to the Lower Delaware Valley had several objectives, which are stated here without any implication of relative importance.

One objective was to be able to say something about the practical importance and difficulty of including within a single model airborne, waterborne, and solid residuals. The regional residuals management problem cannot, in principle, be solved for air or water or solids in isolation because of the links among forms of residuals and discharge media implied by the conservation of mass and energy in production, use, and residuals modification processes. There was, however, no hard evidence on the quantitative extent of the linkages and the size of the costs implied by isolated solutions in real situations. The applied model was meant to be a good enough representation of a real region to yield one piece of defensible evidence, at least on the question of linkages.

A second objective was to generate information on the implications of various REQM strategies within a regional context. For example: What difference does it make if there is an institution available that can construct and operate regional facilities of various types, such as regional sewage treatment plants and in-stream aeration facilities? What effect is there on solid residuals management costs if separate collection of some types of paper residuals with subsequent recycling is included as an option?

A third objective was to explore the computational problems inherent in large-scale regional applications, i.e., how best to deal with large amounts of data, how to aggregate with minimal loss of useful information, how best to decompose large-scale models so that they become computationally tractable.

A fourth objective was to produce an application with sufficient reality to see whether and how such models might work in an actual legislative-executive setting. Experimental work in this area was encouraging, but also indicated that the inclusion of the necessary information on the distribution of costs and benefits added significantly to the difficulties of constructing an applied regional REQM model and promised that in a real setting this information would add significantly to model size. Making the extension would permit an exploration of the degree to which the utility of the regional economic-engineering-ecologic model might be increased by the use of a political model which would allow potential legislative and executive users to explore distributional aspects efficiently, or would at least permit an exploration of the analytical costs of providing politically useful information on the distribution of costs and benefits of REQM strategies.

In the "real world," the range of possible REQM strategies, because of the multiple sources and many types of residuals and the varied distributions of impacts and costs, results in an almost infinite number of possible combinations for analysis. How can such complexity be represented, and the results of analyses presented, to make meaningful decisions possible? Developing data which would shed light on this question in a real situation was another objective, closely related to the fourth.

A final objective was to investigate the costs and benefits implied by incorporating a nonlinear model of the natural world within a regional REQM model optimization framework. It was hoped to arrive at some basis for judging whether the increased computational costs and analytical problems appear to be outweighed by some combination of greater accuracy and increased output of policy-relevant information.

The Region

The region chosen for the application was the Lower Delaware Valley region. This is a complex region with many individual point and nonpoint sources of residuals discharges. It was chosen in preference to simpler regions, where fewer data might be required, for three reasons. First, its very complexity provided a hard test of the applicability of the approach. If it could be applied here, it should be workable almost anywhere. In addition, it could perhaps supply the kernel of a useful analytical tool for REQM policy making in one of the most important regions in the United States. Second, ties existed between Resources for the Future and certain governmental institutions in the region, particularly the Delaware River Basin Commission. Third, this region had been extensively studied for other purposes, so that a body of pertinent information was available surpassing that for most other regions. Accordingly, even though the Lower Delaware Valley is very complex, it promised to offer enough advantages in terms of data availability to offset the costs of analyzing such a complex region.

The region selected is defined by county boundaries and is shown in figure 3-1. It covers an area of about 4,700 square miles (approximately 12,200 square kilometers). The grid superimposed on the figure is used for locating dischargers and receptors of gaseous residuals in the model. It is related to the Universal Transverse Mercator Grid covering the United States. The region consists of Bucks, Montgomery, Chester, Delaware, and Philadelphia Counties in Pennsylvania; Mercer, Burlington, Camden, Gloucester, and Salem Counties in New Jersey; and New Castle County in Delaware.

Figure 3-1. Lower Delaware Valley Region

Note: The grid is in kilometers and is based on the Universal Transverse Mercator (UTM) Grid System.

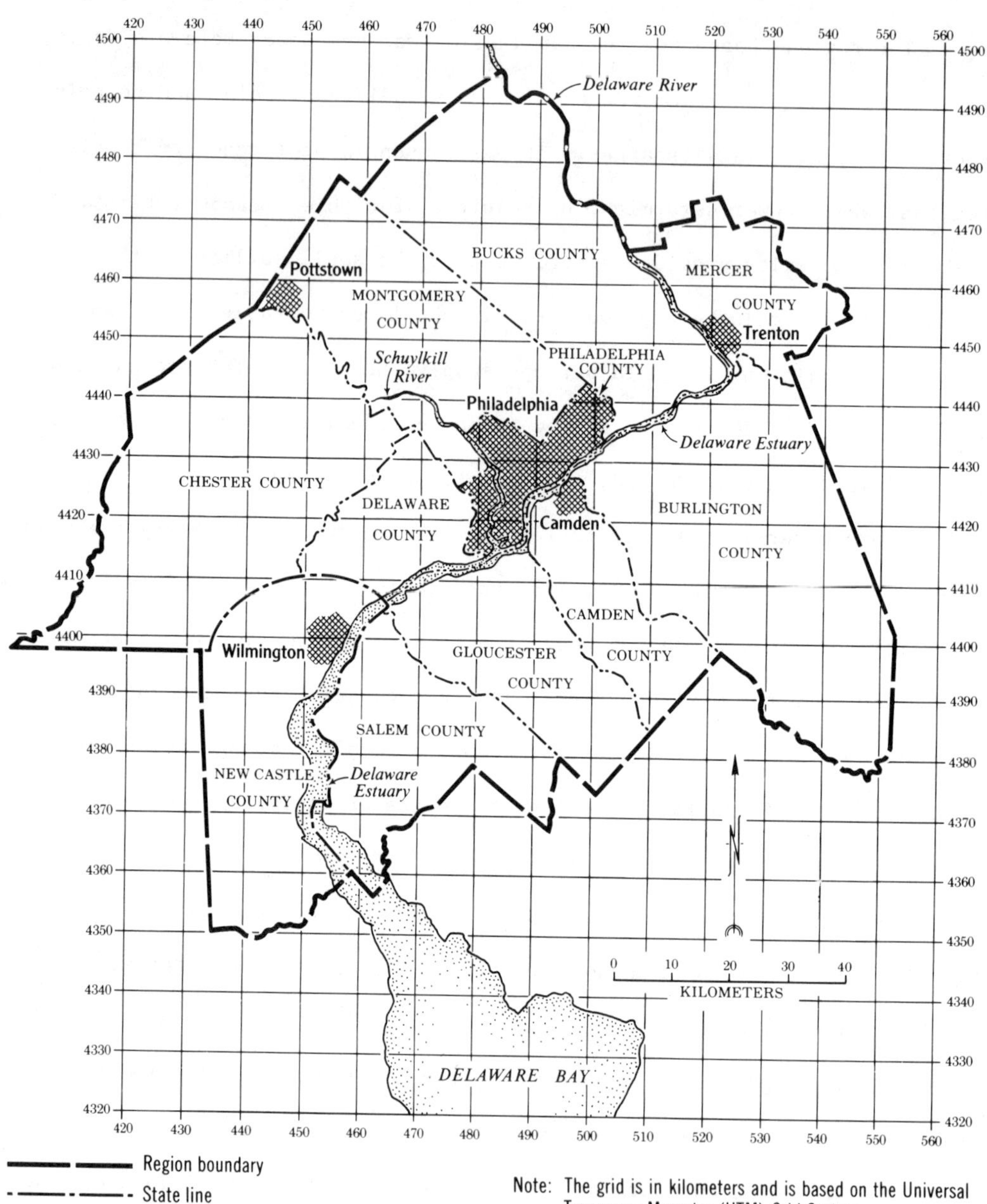

Source: Walter O. Spofford, Jr., Clifford S. Russell, and Robert A. Kelly, Environmental Quality Management: An Application to the Lower Delaware Valley (Washington, D.C.: Resources for the Future, 1976).

The major cities in the area are Philadelphia (coterminous with Philadelphia County) in Pennsylvania; Trenton in Mercer County; Camden in Camden County; and Wilmington in New Castle County. There are 379 incorporated political jurisdictions--cities, towns, townships, boroughs, and divisions--in the eleven counties.

The 1970 population of the eleven counties was a little more than 5.5 million. Of this, 35 percent was accounted for by Philadelphia alone, with a further 5 percent found in Trenton, Camden, and Wilmington. But even these figures do not give a full idea of the extent of urbanization, for 10 of the 11 counties--Salem being the exception--are over 70 percent urbanized.[2] Population densities range from 165 persons per square mile (Salem) to 15,000 (Philadelphia), and are 1,000 or greater per square mile for four other counties. In both Pennsylvania and Delaware, densities in the region are uniformly greater than the average for the respective states, but in New Jersey the "suburban" counties of Burlington and Gloucester and rural Salem County are all less densely populated than the average for the state.

Incomes are generally high in the region. Using median family income as the indicator, every county except Philadelphia has higher median family income than the United States as a whole. Salem County has the lowest median income outside Philadelphia; Montgomery County has the region's highest median income. The range between these two extremes is about $2,500 (per family per year), or about 25 percent of the lower figure. Thus, there are significant intraregional income differences, even at the highly aggregated level of counties. These differences are quite significant for REQM, because experience indicates interest in a cleaner ambient environment is positively correlated with income. They also suggest that the distributional implications of REQM policy will be an important issue in the region. This presumption is underlined by the fact that it is Philadelphia which sits in the center of the largest industrial concentration, has the highest population density, and hence can be expected to have the greatest environmental quality problems, and which also has the lowest per capita income base from which to pay for improving AEQ.

The region is one of the most heavily industrialized areas in the United States. It contains: seven large petroleum refineries; five large steel mills and many smaller ones; thirteen pulp and paper or paper mills with more than 100 tons per day output, and numerous smaller ones; fifteen large thermal electric energy generating plants and two smaller ones; many large and small chemical and petrochemical plants of various types; and many other types of industrial operations, including foundries and automobile and electronic assembly plants.

In 1968, there were seventeen municipal sewage treatment plants in the region with flows greater than one million gallons per day (mgd) discharging to the Delaware Estuary. (Seven of these plants had flows greater than 10 mgd.) In addition, 123 sewage treatment plants of various sizes discharged to the Schuylkill River and to the tributaries of the two major rivers.[3] In 1970 there were seventeen municipal incinerators and numerous landfills and dumps in operation in the region.

The major recipient of waterborne residuals in the region is the Delaware Estuary itself. The estuary is generally defined as the approximately 85-mile stretch of river between the head of the tide at Trenton and the head of Delaware Bay at Liston Point, Delaware. Figure 3-2 shows the estuary and the 22 reaches into which it was divided for analyzing REQM strategies.

Figure 3-2. Reach Locations for the Delaware Estuary Model

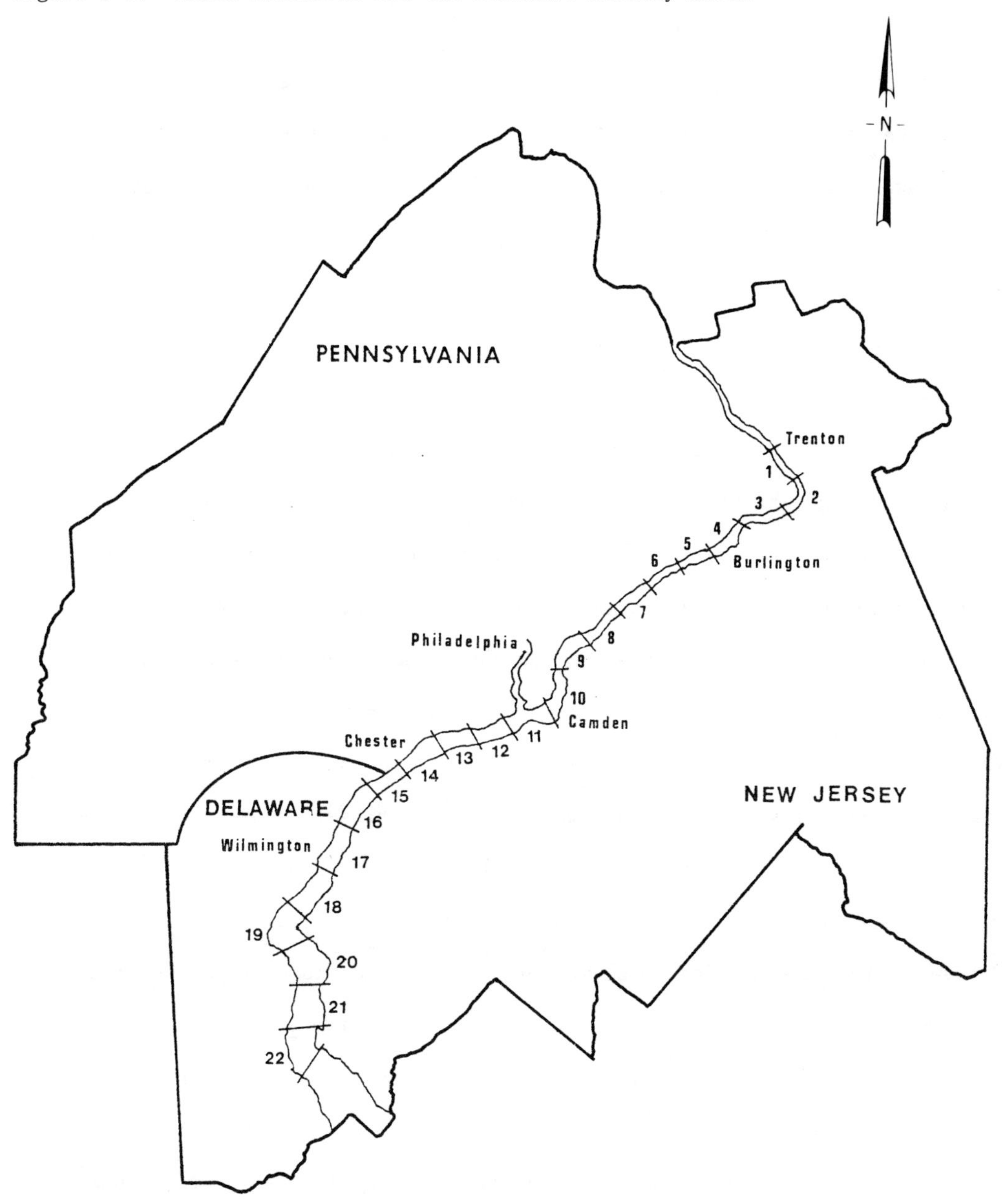

Source: Robert A. Kelly, "Conceptual Ecological Model of the Delaware Estuary," in B.C. Patten, ed., Systems Analysis and Simulation in Ecology, vol. IV (New York: Academic Press 1976).

The flow of the river varies widely, from month to month, and year to year. The low flow period, and hence the period during which the river is most sensitive to residuals discharges, is July-October. The flow of the Delaware River at Trenton, used as the basis for the aquatic ecosystem model, corresponds to the mean flow for September 1970, i.e., about 4,150 cfs. This flow is about 2.8 times as large as the seven-day, ten-year low flow at Trenton of 1,500 cfs. Table 3-1 shows the estimated residuals discharges to the 22 reaches of the estuary for September 1970, and the heavy concentration of those discharges in a relatively few reaches. The maximum and minimum concentrations in the estuary for that month are also shown. The lowest water quality is found in a stretch of the estuary downstream from Philadelphia near the confluence with the Schuylkill River.

The atmospheric "resource" of the region does not lend itself to as simple a characterization as the flow relations for its watercourses. For the region as a whole, the seasonal prevailing wind pattern is roughly: winter and spring westerlies (and west-northwesterlies); summer southwesterlies; and autumn variability. In almost every month there are southwesterly winds along the stretch from roughly the Delaware line to Camden, New Jersey. The net effect of these meteorological conditions, together with the spatial pattern of discharges of gaseous residuals, is a "mountain" of poor air quality along the estuary from Wilmington to Trenton, with the highest concentrations over the Philadelphia-Camden area. On average, the best meteorological conditions for the dilution and dispersion of gaseous residuals in the Philadelphia area appear to be in June, the worst in January.

During the 1967-1968 period the maximum annual average concentrations of sulphur dioxide (SO_2) and suspended particulates (TSP) in the region were

Table 3-1. Estimated Residuals Discharges to the Delaware Estuary, September 1970

(1,000 pounds per day and percentage of total)

	Residual					
	BOD_5		N		P	
Industrial	385	(41%)	117	(53%)	7.9	(15%)
Municipal	425	(46%)	63.8	(29%)	27.0	(50%)
Tributaries	50	(5%)	19.1	(9%)	13.0	(24%)
Storm water	76	(8%)	20.5	(9%)	5.8	(11%)
Total (rounded)	936	(100%)	220	(100%)	53.7	(100%)

Distribution of September 1970 loads:

83% of BOD_5 load is discharged in 27% of reaches

79% of N load is discharged into 27% of reaches

80% of P load is discharged into 27% of reaches

Measured ambient water quality resulting from September 1970 loads:

Minimum mean dissolved oxygen concentration:	1.2 mg/ℓ
Maximum mean BOD_5 concentration:	6.4 mg/ℓ
Maximum mean P (as P) concentration	0.33 mg/ℓ
Maximum mean N (as TKN)[a] concentration	2.5 mg/ℓ

[a] NH_3 + organic N

about 190μgms/m^3 and about 150μgms/m^3, respectively, measured at a station in Philadelphia. These compare with the U.S. primary annual average ambient standards of 80μgms/m^3 and 75μgms/m^3, respectively. Table 3-2 shows the estimated discharges of SO_2 and particulates in the Lower Delaware Valley Region for 1970.

The major dischargers of SO_2 and particulates in the region are the petroleum refineries, steel mills, and power plants. Collectively, their SO_2 discharges amounted to about 1,760 tons per day, roughly 60 percent of the total of about 3,000 tons per day. Their collective discharges of particulates were about 250 tons per day, roughly 40 percent of the total of about 600 tons per day for the region. The power plants in the region clearly represent the largest set of dischargers of both SO_2 and particulates.

Area sources of SO_2 and particulate discharges accounted for about 25 percent and 35 percent of the totals for the region, respectively. However, these sources typically discharge close to the ground and hence contribute proportionally more to ground-level ambient concentrations than the relative magnitudes of their discharges would indicate. "Area sources" of gaseous residuals are not tied to specific stack locations but are treated effectively as though their discharges were uniformly generated over the different subareas identified.

For modelling air quality, the atmospheric conditions used represent the annual joint probability distribution of wind speed, wind direction, and stability conditions for 1970, assumed to be uniform throughout the region.[4] For neither air nor water quality analyses were conditions representing rare events used in the analyses. Ideally, explicit attention would also have been given to this aspect of regional REQM.

Table 3-2. Estimated Sulphur Dioxide and Particulate Discharges in the Lower Delaware Valley region, 1970[a]

(annual average in tons per day)

Discharger Category	Sulphur dioxide	Particulates
Point sources (1,031 stacks)		
Petroleum refineries	410	66
Steel mills	19	58
Power plants	1,332	126
Other point sources	439	133
Subtotal	2,200	383
Area sources (240 areas)		
Home heating[b]	214	25
Other area sources	550	191
Subtotal	764	216
Totals, all sources	2,964	599

a Based on EPA's 1970 inventory of gaseous discharges for the Metropolitan Philadelphia Interstate Air Quality Control Region, supplied by the Division of Applied Technology, Office of Air Programs, U.S. Environmental Protection Agency, Durham, North Carolina.

b Estimated from the number of housing units and the home heating fuel types contained in the 1970 Bureau of the Census computer tapes.

The Residuals Management Model

The model framework is shown in figure 3-3. There are three main parts of this model: a linear programming model of residuals generation and discharge (comprising both production and use activities); the environmental models; and the environmental evaluation section. The model is designed to provide the minimum cost way of: producing an exogenously determined "bill of goods" at the individual industrial plants; meeting electricity and home and commercial space heating requirements for the region; and handling, modifying, and disposing of specified quantities of municipal liquid and solid residuals, subject to constraints on:

(1) the distribution of environmental quality--water, air, and landfills--over geographic units; and

(2) the distribution of consumer costs--electricity, heating fuel, sewage disposal, solid wastes disposal, regional in-stream aeration--over political jurisdictions.

The former types of constraints are represented by, for example, maximum concentrations of SO_2 and TSP at a number of receptor locations in the region; minimum concentrations of dissolved oxygen and fish biomass in the estuary; maximum concentration of algae in the estuary; and restrictions on the types of landfill operations which can be used in the region. With respect to the latter category of constraints, the model permits constraining increases in: the cost of electricity due to all required discharge reduction activities (by utility service area); the cost of home heating due to fuel switching; and municipal expenditures due to increased liquid residuals modification and more expensive solid residuals management methods. However, credit is given for the sale of newsprint and linerboard produced from used

Figure 3-3. Schematic Diagram of the Lower Delaware Valley Residuals-Environmental Quality Management Model

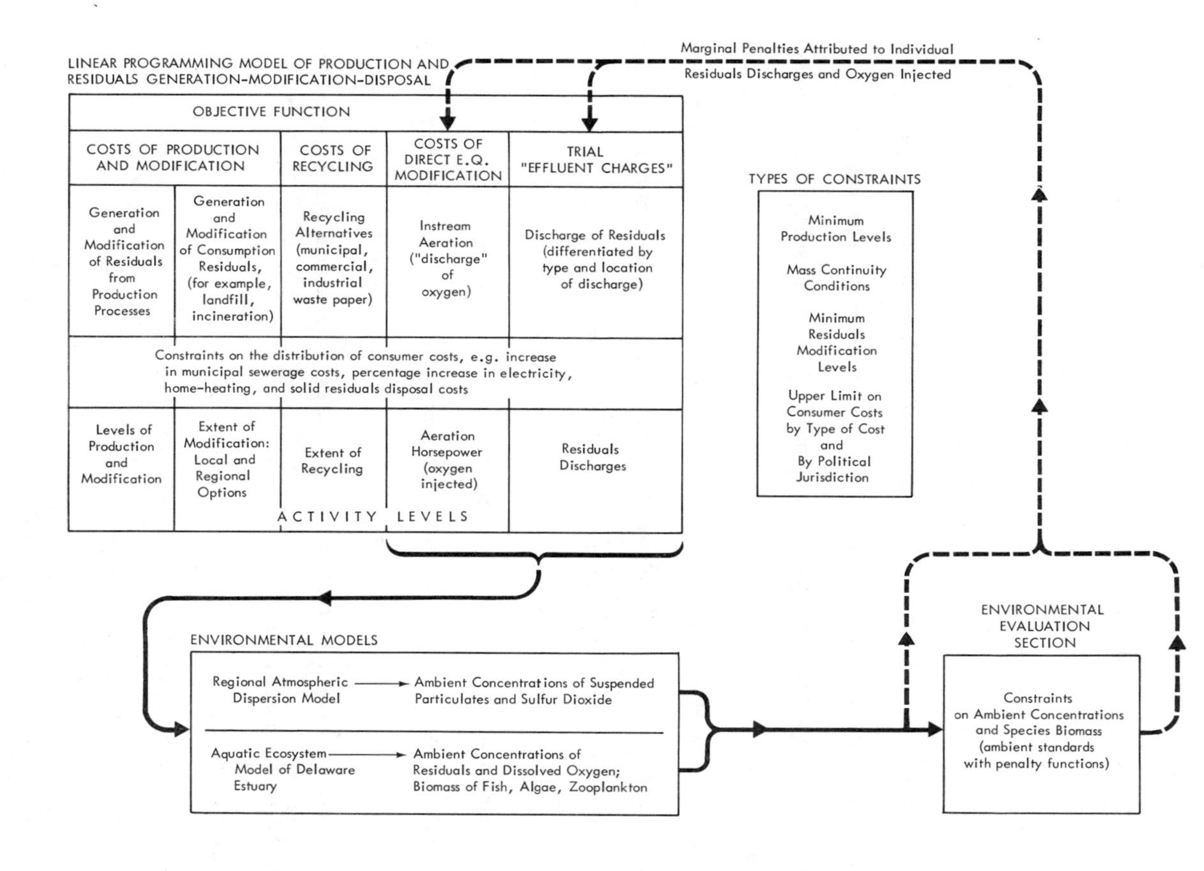

Source: Walter O. Spofford, Jr., Clifford S. Russell, and Robert A. Kelly, <u>Environmental Quality Management: An Application to the Lower Delaware Valley</u> (Washington, D.C.: Resources for the Future, 1976).

newspapers and used corrugated containers, respectively, collected within the region, such production being an alternative to solid residuals disposal to landfills and/or incineration.

Referring to figure 3-3, in the upper left-hand block of the diagram is found the basic driving force for the entire model, a linear programming model of residuals generation and discharge. It is in this part of the model that the minimum "production" constraints are found. A key output of this part is a vector of residuals discharges, identified by substance and location. These feed into the environmental models: the model of the aquatic ecosystem and the dispersion model for SO_2 and TSP. This section of the overall model, in turn, produces as output a vector of AEQ levels (for example, SO_2 in the 37th political jurisdiction). These concentrations are then treated as input to the "evaluation" submodel found in the lower right of the diagram. Here the concentrations implied by one solution of the residuals generation and discharge submodel are compared with the exogenously established environmental standards imposed for the model run. An iterative procedure, based on the gradient method of nonlinear programming and on the use of penalty functions, was developed to meet the ambient standards within some specified tolerance, and to select a vector of residuals discharges that meets the constraints at least cost to the region.

The following are the main features of the model.

(1) It is an optimization model, in contrast to a simulation model. It is intended to provide least-cost ways of achieving various levels of AEQ subject to constraints on production and use, consumer costs, and AEQ.

(2) It reflects non-treatment alternatives available for reducing the amounts of residuals initially generated, especially in production processes. The amount, or form, of residual to be handled is not, in general, assumed to be fixed.

(3) It deals with the three major forms of materials (liquids, gases, and solids) and the three receiving environmental media (watercourses, atmosphere, and land) simultaneously. The model reflects the conservation of mass and energy for relevant residual components, accounting for material and energy flows as they are modified from one form to another in production, use, and residuals modification activities. Carbon dioxide and water vapor are ignored.

(4) It is capable of incorporating various types of models of the natural world from the simplest linear transformations, for example, the steady-state Streeter-Phelps type dissolved oxygen models, to complex simulation models of aquatic ecosystems.

(5) It is a static, economic model so that time is not considered in the residuals generation and discharge portion of the model; capacity expansion and optimal timing are not explored; no new industries come into the region; the population level and its distribution remain constant. The model output represents a "snap shot" of the region for fixed conditions at a given point in time; a dynamic analysis of the impact on the region of economic and population growth cannot be made, but a comparative static analysis could be made.

(6) The spatial distribution of activities in the region is fixed, although the impacts of various distributions could be examined and compared.

(7) Production and use activities in the region are deterministic and steady state.

(8) The natural world models are deterministic and steady state, although the ecosystem model is structured as a nonsteady-state model, but the eventual steady-state results are used in the calculations.

(9) A single season (spring, summer, fall, winter, annual) is employed for the analysis, although the model could be operated for different seasons or time periods and the optimal solutions for each season compared.

(10) Interactions among residuals in the atmosphere do not occur, and decay rates, where applicable, are independent of the quantities of the residual present and of the presence of other residuals.

To provide specific information on the geographic distribution of AEQ and of REQM costs to consumers throughout the region, the Lower Delaware Valley was divided into 57 political jurisdictions with 1970 populations ranging from about 60,000 to about 110,000. To form these jurisdictions, some of the 379 cities, towns, boroughs, and townships that comprise this region were aggregated and others were subdivided. However, all 57 political jurisdictions are comprised of whole census tracts and are located entirely within the boundaries of individual counties and hence within states. For each political jurisdiction the following measures of AEQ and of increased consumer costs resulting from REQM are available: air quality--annual average ground-level concentrations of SO_2 and TSP at an arbitrarily selected location within each jurisdiction; landfill quality; and levels of increased consumer costs (uniform over each jurisdiction) for electrical energy, home and commercial heating, municipal sewage handling-modification-disposal, and municipal solid residuals disposal. Ambient water quality in the estuary--in terms of concentration of dissolved oxygen, and biomass of fish and algae--is available for each of the 22 reaches shown in figure 3-2 and is related to the 57 political jurisdictions only indirectly. The allowable upper, or lower if appropriate, limits of each of these measures can be constrained.

The estuary is the only water body in the region which is described by a water quality model. Dischargers in the reach of the Schuylkill River included in the model are handled with specified discharge standards based approximately on state standards existing in 1968-70. This arrangement does not permit liquid discharge trade-offs in this part of the region. Dischargers to other tributaries of the Delaware estuary are dealt with in the same manner as dischargers to the Schuylkill. The stretch of the Delaware River above Trenton currently is of fairly good quality, and there is little prospect for major deterioration in the near future. Consequently, no water quality model of this part of the region was considered necessary.

Residuals Generation and Discharge Models

The residuals generation and discharge portion of the Lower Delaware Valley model is composed of six separate linear programming (LP) models arranged in individual modules. These modules are depicted in table 3-3. The first column indicates the module number, for identification purposes only. (The MPSX designation derives from the particular computational routine used in the analysis.) The next three columns indicate the sizes of the individual LP modules--number of rows, number of columns, and number of residuals discharges, respectively. Note that in total in the model there are: almost 8,000 variables (columns); a little over 3,000 constraining relationships (rows); and almost 800 individual residuals discharges that enter the environmental models. The residuals involved are BOD_5, nitrogen, phosphorus, toxics (phenols), suspended solids, and heat discharges to the estuary, and SO_2 and particulate discharges to the atmosphere. The fifth column describes the type and indicates the number (in parentheses) of activities in the region for which residuals management options are included in the model. The sixth column lists the extra cost distributional constraints available in the model. Except for the costs of sewage disposal, there is one constraint for each type of extra cost and for each of the fifty-seven political jurisdictions into which the Lower Delware Valley has been divided for modeling purposes. For sewage disposal, there are only forty-six extra costs in the model because eleven jurisdictions do not discharge any sewage directly to the estuary.

In the model, 297 point and nonpoint sources of gaseous and liquid residuals were provided with management options for reducing their discharges. The 183 point sources include 124 industrial plants and fifty-nine municipal

Table 3-3. Lower Delaware Valley Model:
Residuals Generation and Discharge Modules*

Module identification	Size of linear program			Description	Percent extra costs for the 57 political jurisdictions (except as noted)
	Rows	Columns	Discharges		
MPSX 1	286	1,649	130	Petroleum refineries (7) Steel mills (5) Power Plants (17)	57 electricity
MPSX 2	741	1,482	114	Home heat (57) Commercial heat (57)	57 fuel 57 fuel[d]
MPSX 3	564	1,854	157	"Over 25 $\mu gms/m^3$" dischargers (75)[b]	
MPSX 4	468	570	180	Delaware Estuary sewage treatment plants (36)	46 sewage disposal ($ per household per year)[e]
MPSX 5	951	1,914	88	Paper plants (10) Municipal incinerators (23) Municipal solid residuals handling and disposal activities	57 solid residuals disposal (incinerators and solid residuals activities)
MPSX 6	229	395	117[a]	Delaware Estuary industrial dischargers (23)[c] In-stream aeration (22)	57 in-stream aeration (absolute extra cost per day)[f]
TOTALS	3,239	7,864	786		

*See next page for footnotes.

Footnotes to Table 3-3:

[a]Does not include the 22 oxygen "discharges" from the in-stream aeration option.

[b]Includes each industrial plant whose gaseous residuals discharges result in maximum annual average ground level concentrations equal to or greater than 25 μgms/m^3. To determine this group, all stacks in the IPP inventory were considered except those in MPSX 1. The maximum annual average ground level concentrations of SO_2 and TSP were computed for each stack. For all stacks at the same x-y location (i.e., same plant), the maximum ground level concentrations were added together. Those plants resulting in maximum ground level concentrations, for either SO_2 or TSP, equal to or greater than 25 μgms/m^3 were then placed in this category.

[c]Twelve of the Delaware Estuary industrial wastewater dischargers in MPSX 6 are also represented by SO_2 and/or suspended solids dischargers in MPSX 3, and the gaseous residuals discharges of another are included in MPSX 5.

[d]The commercial heating requirements in this module are based on the differences between SO_2 discharges from area sources in the IPP inventory of gaseous discharges and SO_2 discharges from the home heating model. Commercial heating requirements for political jurisdictions 19, 28, 29, 31, 32, 33, 37, and 44 are equal to zero.

[e]The extra costs reported represent the average increase per household per year for each political jurisdiction. Jurisdictions 5, 6, 7, 36, 39, 43, 45, 46, 53, 54, and 56 do not discharge to the estuary at all.

[f]The model currently reports the total regional absolute extra cost per day for in-stream aeration. The cost is then allocated equally among the 57 political jurisdictions. Any other distribution is also possible.

Note: The major sources of information and data used in the construction of the residuals generation and discharge models were: (i) EPA's 1970 inventory of SO_2 and particulate discharges in the Metropolitan Philadelphia Interstate Air Quality Control Region; (ii) Delaware River Basin Commission for the municipal and industrial wastewater dischargers to the Delaware Estuary; (iii) Regional Conference of Elected Officials publications for municipal incinerators and for electricity districts and power plants; and (iv) the 1970 Bureau of the Census computer tapes for number of people and housing units by political jurisdictions, home heating fuel characteristics, and sewerage systems. The base period for data employed in this model is 1968 to 1970.

residuals handling and disposal activities--twenty-three incinerators and thirty-six sewage treatment plants. The 124 industrial plants in turn include, among others, the seven oil refineries in the region, five steel plants, seventeen power plants, and twenty-three other industrial dischargers to the estuary. The 114 area (nonpoint) sources are 57 home heating activities and 57 commercial heating activities (one of each for each jurisdiction). The other point and nonpoint sources identified in the region that are not provided with residuals management options are incorporated as background dischargers.

The management options available to the various sources for reducing their discharges range from alternative production processes and raw material inputs to residuals modification at the "end of the pipe." Examples of these options are shown in table 3-4. In addition to the management options for the individual activities listed above, four collective residuals handling and modification activities, and one direct modification of AEQ activity are included: incineration, landfilling, regional sewage modification, and recycling of used newspapers and used corrugated containers represent the former; in-stream aeration, the latter.

The sources of residuals discharges included in the regional model do not encompass all residuals dischargers in the Lower Delaware Valley region. Certain kinds of activities, and residuals, have not been included, such as transportation and agricultural activities and land runoff, except as "background" sources of residuals where appropriate. For certain kinds of AEQ problems, these sources are important. The activities that have been included in the model, however, include the important dischargers of SO_2 and particulates to the atmosphere and the major dischargers of BOD_5, nitrogen, and phosphorus to the Delaware Estuary.

Table 3-4. Examples of Residuals Management Options Available to Selected Types of Dischargers in the Lower Delaware Valley Model

Type of Residual Discharger	Management Option [b] Available	Primary Residual Reduced	Secondary Residual [a] Generated
Petroleum Refineries	1. Charge lower sulphur crude (2 grades: 0.4, 1.44% sulphur)	SO_2	None
	2. Sell, rather than burn, certain high-sulphur fuel products, such as 3.3% sulphur refinery coke	SO_2 and Particulates	None
	3. Use purchased lower sulphur fuel (3 sulphur contents: 0.5, 1.0, 2.0%)	SO_2	None
	4. Cooling tower(s) for segregated, noncontact cooling water	Heat	c
	5. Cyclone collectors on cat-cracker catalyst regenerator (2 efficiencies: 70, 85%)	Particulates	Fly ash
	6. Electrostatic precipitator on cat-cracker catalyst regenerator, 95% efficiency	Particulates	Fly ash
	7. Sludge drying and incineration	Sludge	Particulates, Suspended solids
Thermal Power Generating Facilities	1. Use lower sulphur fuel		
	a. Coal (4 sulphur contents: 0.5, 1.0, 1.5, 2.0%, with associated ash contents of 7.4, 8.1, 8.8, and 9.5% respectively)	SO_2 and Particulates	None
	b. Residual fuel oil (4 sulphur contents: 0.5, 1.0, 1.5, 2.0%)	SO_2 and Particulates	None
	c. Natural gas		
	2. Cooling tower(s)	Heat	c
	3. Limestone injection-wet scrubber, 90% efficiency	SO_2 and Particulates	Lime slurry
	4. Electrostatic precipitator (3 efficiencies: 90, 95, 98%)	Particulates	Fly ash
Commercial Space Heating	1. Use lower sulphur fuel		None
	a. Residual fuel oil (sulphur contents: 0.5, 1.0, 1.5, 2.0%)	SO_2	
	b. Natural gas	SO_2 and Particulates	None
Municipal Sewage Treatment Plants	1. Wastewater treatment (2 efficiencies: BOD_5 77, SS 80; BOD_5 97, SS 99%)	BOD_5 and Suspended Solids	Sludge
	2. Sludge digestion, drying and landfill	Sludge	d
	3. Sludge dewatering and incineration	Sludge	Particulates, Fly ash
	4. Dry cyclone on sludge incinerator (2 efficiencies: 40, 96%)	Particulates	Bottom ash

[a] A secondary residual is one generated in the process of modifying a primary residual.
[b] Efficiency refers to removal efficiency.
[c] Heat is rejected to the atmosphere along with water vapor.
[d] The secondary residual is digested sludge at a different location.

Environmental Models

The overall regional REQM model incorporates a nonlinear ecosystem model of the Delaware Estuary, divided into twenty-two reaches, as shown in figure 3-2. Inputs of liquid residuals discharges to this model are: organics (BOD_5), nitrogen, phosphorus, toxics (phenols), suspended solids, and heat (Btu). Outputs are expressed in terms of ambient concentrations of algae, bacteria, zooplankton, resident fish, dissolved oxygen, BOD_5, nitrogen, phosphorus, toxics, suspended solids, and temperature. Three of these outputs--algae, fish, and oxygen--can be constrained. These constraint levels are set exogenously to the management model and represent both a major driving force in the solution runs and one of the principal policy issues the model is designed to study. Thus, the cost of meeting alternative sets of AEQ levels can be viewed as the most important single output capability of the regional model. A summary of the estuary model indicating relevant inputs and outputs is presented in table 3-5, and a diagram showing the structure of the model in terms of materials flows is shown in figure 3-4.

With respect to air quality, the regional management model includes two 57 x 240 (57 receptor locations, one for each political jurisdiction, and 240 dischargers) air dispersion matrices--one each for SO_2 and TSP. These matrices relate annual average ambient ground level concentrations to discharges of SO_2 and particulates. The matrices are based on output of the air dispersion model contained in EPA's Implementation Planning Program. The air dispersion model was calibrated using equations developed in an unpublished EPA study of the Philadelphia Metropolitan Air Quality Control Region. A summary of the air dispersion model, indicating relevant inputs and outputs, is shown in table 3-6.

Table 3-5. Delaware Estuary Ecosystem Model[a]

Endogenous Variables (compartments--mg/ℓ)

Algae
Zooplankton (herbivores, detritivores, and bacterivores)
Bacteria
Fish
Dissolved oxygen (DO)
Organic matter (as BOD_5)
Nitrogen
Phosphorus
Toxics
Suspended Solids
Temperature (°C)

Inputs of Residuals (pounds/day)

Organic material (as BOD_5)
Nitrogen
Phosphorus
Toxics
Suspended solids
Heat (Btu)

Target Outputs[b] (mg/ℓ)

Algae
Fish
Dissolved oxygen

Model

Type: materials balance-trophic level
Characteristics: deterministic, nonsteady-state
Calibration: based on average September 1970 flow at Trenton, N.J., of 4150 cfs.

Reaches

Number: 22

---

[a]For details, see: Robert A. Kelly, "Conceptual Ecological Model of the Delaware Estuary," in B.C. Patten, ed., Systems Analysis and Simulation in Ecology, vol. IV (New York: Academic Press, 1976), and Robert A. Kelly and Walter O. Spofford, Jr., "Application of an Ecosystem Model to Water Quality Management: The Delaware Estuary," in C.A.S. Hall and J.W. Day, Jr., eds., Ecosystem Modeling in Theory and Practice: An Introduction with Case Histories (New York: John Wiley and Sons, Inc. 1977).

[b]The management model is operated for relevant minimum, or maximum, allowable ambient concentrations ("standards").

Figure 3-4. Diagram of materials flows among compartments within a single reach: Delaware Estuary Ecosystem Model

Notation:

N = nitrogen	B = bacteria	L = organic matter (as BOD_5)
P = phosphorus	H = zooplankton	O = dissolved oxygen
A = algae	F = fish	

Note: The three remaining endogenous variables--heat (temperature), toxics, and suspended solids--are assumed to affect the rates of material transfers among the ecosystem components, or "compartments".

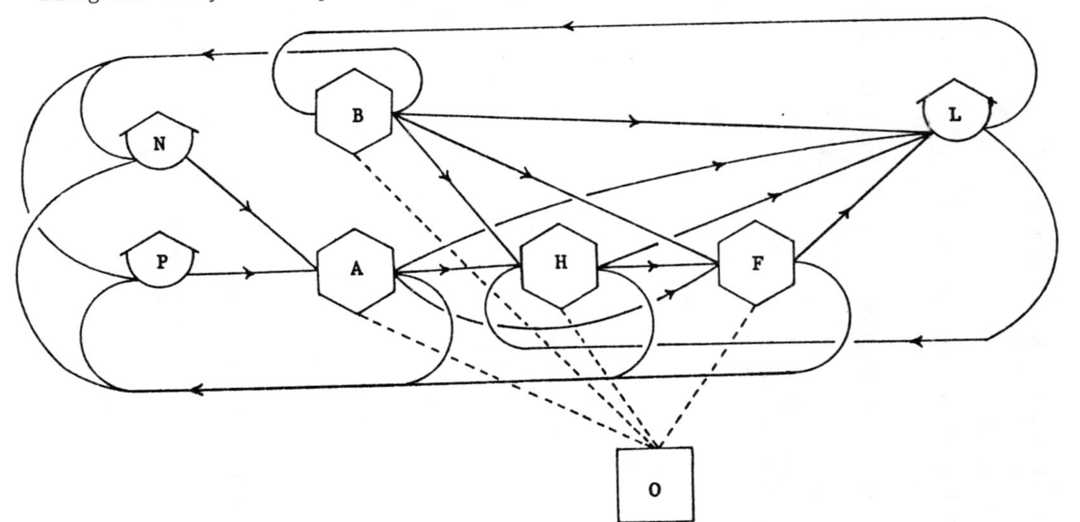

Source: Walter O. Spofford, Jr., Clifford S. Russell, and Robert A. Kelly, "Operational Problems in Large-Scale Residuals Management Models," in Edwin S. Mills, ed., Economic Analysis of Environmental Problems (New York: National Bureau of Economic Research, 1975).

Table 3-6. Air Dispersion Model[a]

Endogenous Variables (ground level, annual average ambient concentrations-- μ gms/m^3)

Sulphur dioxide
Suspended particulates

Inputs of Residuals (tons/day)

Sulphur dioxide
Particulate matter

Target Outputs[b] (μ gms/m^3)

Sulphur dioxide
Suspended particulates

Model

Type: Gaussian plume dispersion model from EPA's Air Quality Implementation Planning Program (IPP)
Characteristics: deterministic, steady-state

Input data requirements

1. Sources

(a) x-y coordinates of each stack
(b) discharge rates for each stack
(c) physical stack height
(d) stack diameter
(e) stack exit temperature
(f) stack exit velocity

2. Receptors

x-y coordinates of each receptor location

3. Meteorological Conditions

(a) annual joint probability distribution for wind speed (6 classes), wind direction (16 directions), and atmospheric stability (5 classes), using 1970 data for Philadelphia; the result is 480 discrete meteorological situations

(b) mean temperature: 68°F (20°C)[c]

(c) mean annual pressure: 1,017 millibars (30.03 inches of mercury)

(d) mean maximum afternoon atmospheric mixing-depth: 1,000 meters[d]

---

a For details, see TRW, Inc., "Air Quality Implementation Planning Program," U.S. Environmental Protection Agency, vols. I and II, November 1970 (also available from National Technical Information Service, Springfield, VA 22161, as nos. PB-198 299 and PB-198 300, respectively).

b The management model is operated for maximum allowable ambient concentrations ("standards").

c The mean temperature used in the model, 68°F, is roughly that for September. The mean annual temperature is 64°F.

d The mean maximum afternoon mixing depth used in the model, 1,000 meter is the average for the months of May, August, and September. In June and July, the mean mixing depth is greater than 1,000 meters; from October to April, it is less than 1,000 meters. The effect of decreasing the mixing depth around the 1,000 meter level is to increase the concentrations predicted by the model.

To keep the two air dispersion matrixes as small, and thus as manageable, as possible, all stacks at the same plant location in EPA's inventory of gaseous residuals discharges were aggregated to a single stack. This required computing the characteristics of a single stack that produced the same result as the group of stacks, i.e., the same maximum ground-level concentration as the group, using discharge rates from EPA's inventory. The discharge rate employed in the analysis for the single stack was the aggregate discharge of all the individual stacks.

Stack characteristics of the single stack (physical stack height, plume rise, and effective stack height) were determined for discharges of both SO_2 and particulates. In general, the two "artificial" stacks computed on the basis of the above criterion differed substantially because the characteristics of the single stack depend on the relative discharge rates among the individual stacks. In developing the two air dispersion matrices, two different sets of single stacks were used, one for the SO_2 analysis, another for the particulates analysis.

Setting Up the Model Runs

The operational objective in running the model was to explore the effects on aggregate regional REQM costs and the distribution of those costs of the following decision variables: different requirements on the quality of landfills; different ambient water and ambient air quality standards; an in-stream aeration option for directly improving estuary water quality; a regional sewage treatment plant option; and a regional solid residuals management option. For both air and water quality, there are two sets of standards, a relatively easy (E) set to meet and a stricter (T) set. For landfill operations, three quality levels are specified: low, medium, and high. The sets of standards for the three media, which sets are used in various combinations in the analyses, are shown in table 3-7.

Analysis with the model involves specifying for each production run a combination of environmental quality standards to be met and permitted regional residuals handling and modification options. Most runs were arbitrarily stopped at between 29 and 35 iterations, although technically the optimum, even a local optimum, was not necessarily achieved.[5] However, experience indicated that: at this number of iterations the environmental constraints are met, if it is technically possible to meet them; this location is high up on the response surface; and continuing beyond this point improves the objective function value only slightly. This is not meant to imply, however, that residuals discharge patterns would not change, perhaps even substantially, if the run were to be continued beyond 30 iterations. Experience in running the model indicates that, although the total residuals discharges do not change much beyond the 30 to 35 iterations (because the ambient standards are being met), trade-offs in reductions among individual dischargers are still taking place.[6] Thus, a very relevant question is, how comparable are the results--costs and discharges--of the different production runs using output obtained from approximately the 30th iteration of each run.

Results of Running the Model

The combinations of decision variables used in selected runs of the model are indicated in table 3-8 , along with the number of iterations for

Table 3-7. Air, Water, and Land Environmental Quality Standards for Production Runs of the Lower Delaware Valley Model

	Easy to Meet Standards ("E")	More Difficult to Meet Standards ("T")
Water Quality		
Algae	3.0 mg/ℓ	2.0 mg/ℓ
Fish	0.01 mg/ℓ	0.03 mg/ℓ
Dissolved oxygen	3.0 mg/ℓ	5.0 mg/ℓ
Air Quality		
Sulphur dioxide	120 μgms/m^3	80 μgms/m^3*
Suspended particulates	120 μgms/m^3	75 μgms/m^3*

*For certain jurisdictions, it is technically impossible, as the model is constructed, to meet the stricter standard. Consequently, the following modifications were made to the stricter standards:

Political jurisdiction	TSP	Political jurisdiction	SO_2
14	76 μgms/m^3	16	83 μgms/m^3
15	82		
22	76		
23	79		
24	79		
27	81		
50	76		

Landfill Quality

L (low quality) -- Open dump, but no burning allowed.

M (medium quality) -- Good quality sanitary landfill.

H (high quality) -- Good quality sanitary landfill with shredding, impervious layer to protect groundwater, treatment of leachate, aesthetic considerations such as fences, trees, etc.

Table 3-8. Characterization of Selected Production Runs of the Lower Delaware Valley Model and Resulting Values of the Objective Function *

Run Number	Standards[a]			Regional options[b]	Iterations of Regional Model	Objective Function Value[c] (10^6 $/day)
	Air	Water	Landfill	In-stream aeration		
1	T	T	H	+	34	1.20
2	T	E	H	+	35	1.18
3	E	T	H	+	29	0.42
4	E	E	H	+	30	0.36
5	T	T	M	+	35	1.16
6	T	E	M	+	34	1.14
7	E	T	M	+	30	0.37
8	E	E	M	+	30	0.31
9	T	0	H	+	35	1.12
10	0	E	H	+	30	0.11
11	0	T	H	+	16	0.14
12	E	0	H	+	30	0.26
13	E	T	H	0	30	0.52
14	E	E	H	0	30	0.35
15	E	T	H	0[d]	29	0.46
16	E	T	H	+[d]	30	0.39
17[e]	∞	0	H	0	1	1.19
18[e]	0	∞	H	0	1	0.34
19[e]	∞	∞	H	0	1	1.41
23	0	∞	H	+	1	0.36
25	0	0	H	0	1	0.034
26	0	0	M	0	1	0.016
Base Case	0	0	L	0	1	0

* See next page for footnotes.

Footnotes to Table 3-8

a Notation for standards (See table 3-7)

Air and water quality

T = most restrictive standards

E = easier to meet standards

0 = no effective standard

∞ = extremely high marginal penalties on residuals discharges

Landfill quality:

H = high quality landfill

M = medium quality landfill

L = low quality landfill

b The regional options include: two regional paper recycling plants (a newsprint plant and a linerboard plant); two regional sewage treatment plants; and instream aeration. The options of using used newspaper as input to the newsprint plant and used corrugated containers as input to the linerboard plant as alternatives for reducing solid residuals disposal costs are available in all computer runs. The regional sewage treatment plant option is available only in runs 15 and 16. + means instream aeration is employed in the run; 0 means instream aeration is not employed.

c Increased costs, i.e., net of base costs.

d Regional sewage treatment plant option also employed.

e Runs 20, 21, 22 are similar to runs 17, 18, and 19 except that medium quality landfills rather than high quality landfills are specified.

each run and the resulting objective function values. The objective function values represent the additional costs (10^6 \$ per day) over base costs, associated with improving the AEQ of the region--atmosphere, estuary, and landfills. The base case is the situation that existed in the Lower Delaware Valley during the period of roughly 1968 to 1970. The base cost was determined using the Lower Delaware Valley REQM model with landfills of low quality throughout the region, and with only the physical restrictions on liquid and gaseous residuals discharges which existed in the 1968-70 period. Thus the objective function value for the base case is zero.

The results of analyses of the cases specified in table 3-8 illustrate: (1) total regional costs to meet various combinations of ambient air, ambient water, and landfill quality standards; (2) the effects of including regional facilities as components of REQM strategies; (3) linkages among forms of residuals and among the three environmental media; and (4) distributions of changes in AEQ and of REQM costs. The first three are illustrated in figures 3-5, 3-6, 3-7, 3-8, and 3-9 and associated text; the last in tables 3-9 and 3-10 and associated text.

Costs and Environmental Quality Standards

Figure 3-5 shows the total costs to the region of meeting: different sets of air quality standards; different sets of water quality standards; and different sets of air and water quality standards simultaneously. The cost data contained in this sixteen-element table are associated with high landfill quality and with combinations of the following air and water quality constraints:

Figure 3-5. Total Increased Costs to the Lower Delaware Valley Region of Meeting Different Sets of Air and Water Quality Standards, with High Quality Landfills

Water Quality Standards	Air Quality Standards: 0	E	T	∞
0	0.034[a] (#25)	0.26 (#12)	1.12 (#9)	1.19[a] (#17)
E	0.11 (#10)	0.36 (#4)	1.18 (#2)	
T	0.14 (#11)	0.42 (#3)	1.20 (#1)	
∞	0.34[a] (#18)			1.41[a] (#19)

Notes: All values are in 10^6 \$/day.

The numbers in parentheses are the production run numbers.

[a]Single iteration runs, the outputs of which are not dependent on the non-linear optimization algorithm. Aerators are not employed in these runs.

O = no monetary penalties or physical constraints on residuals discharges;

E = easy to meet standards (see table 3-7);

T = more difficult to meet standards (see table 3-7);

∞ = monetary penalties on residuals discharges large enough to force the maximum possible reduction in residuals discharges obtainable in the model.

As expected, total costs to the region increase as the air and water quality standards become more stringent. Note that total costs to the region of meeting T-level water quality standards in the absence of any constraints on the discharge of gaseous residuals are substantially less than the costs of meeting T-level air quality standards in the absence of any constraints on the discharge of liquid residuals. This is due in part to the fact that most of the gaseous residuals dischargers in the region have been considered, but only those liquid residuals dischargers have been considered which discharge their waterborne residuals to the Delaware Estuary. Nevertheless, it appears that "cleaning up" the air will be a more pervasive and costly problem for the region than "cleaning up" the water.[7]

Figure 3-6 illustrates the effects of different landfill qualities in conjunction with different sets of ambient air and water quality standards. The total incremental costs to the region of going from medium quality landfills to high quality landfills amount to between $40,000 and $50,000 per day, depending on the levels of air and water quality standards. However, as indicated by the Module 5 costs, only $25,000 to $30,000 per day are attributable to the paper plant-solid residuals management module. Thus, landfill quality requirements impose substantial incremental costs on other, apparently unrelated, activities in the region. A second point to note from figure 3-6

Figure 3-6. Total Increased Costs to the Lower Delaware Valley Region with High and Medium Quality Landfills

Total Regional Costs:

High quality landfills

Water Quality Standards \ Air Quality Standards	E	T
E	0.36 (#4)	1.18 (#2)
T	0.42 (#3)	1.20 (#1)

Medium quality landfills

Water Quality Standards \ Air Quality Standards	E	T
E	0.31 (#8)	1.14 (#6)
T	0.37 (#7)	1.16 (#5)

Difference in costs

Water Quality Standards \ Air Quality Standards	E	T
E	0.05	0.04
T	0.05	0.04

Module 5 costs:[a]

High quality landfills

Water Quality Standards \ Air Quality Standards	E	T
E	0.060 (#4)	0.079 (#2)
T	0.071 (#3)	0.079 (#1)

Medium quality landfills

Water Quality Standards \ Air Quality Standards		T
E	0.038 (#8)	0.054 (#6)
T	0.040 (#7)	0.055 (#5)

Difference in costs

Water Quality Standards \ Air Quality Standards	E	T
E	0.022	0.025
T	0.031	0.024

Notes: All values are in 10^6 \$/day.
Numbers in parentheses are the production run numbers.

[a] Costs of paper production from paper residuals plus costs of other solid residuals handling and disposal.

is that the water quality standards seem to have very little impact on the costs of paper production from paper residuals and of solid residuals management (Module 5). However, the air quality standards appear to have a substantial impact on these costs.

Costs and Regional Facilities

The effects on costs of including regional options of in-stream aeration and regional sewage treatment plants are illustrated by the results of four runs--3, 13, 15, 16. These runs all have T-level water quality standards, E-level air quality standards, and high quality landfills. When neither the in-stream aeration nor regional sewage treatment plant options are included, total regional costs are $520,000 per day (run 13); with regional sewage treatment but no in-stream aeration, total regional costs amount to $460,000 per day (run 15); with in-stream aeration but no regional treatment, costs are $420,000 per day (run 3); and with both regional treatment and in-stream aeration, total regional costs are $390,000 per day (run 16).

The input activity level of the regional newspaper plant illustrates how REQM conditions affect the economic feasibility of recycling waste paper in the region. Figure 3-7 shows the inputs to the new newsprint plant under varying AEQ standards, from which two conclusions can be drawn. First, operation of the new newsprint plant is very sensitive to landfill costs, as would be expected. For low landfill quality coupled with no AEQ standards (base case), the input to the plant is zero. However, for high landfill costs but still no AEQ standards, the input is 550 tons per day (run 25). At this level of activity, substantial amounts of used newspapers are collected from single-family residences throughout the region, mostly in Philadelphia, in addition to the amounts currently being collected for

Figure 3-7. Quantities of Used Newspaper Input into New Newsprint Plant Under Varying Environmental Quality Standards, With High Quality Landfills

Air Quality Standards

Water Quality Standards	O	E	T	∞
O	550 (#25)	500 (#12)	0 (#9)	0 (#17)
E	160 (#10)	155 (#4)	165 (#2)	
T	0 (#11)	0 (#13)	0 (#1)	
∞	0 (#18)			0 (#19)

Notes: All values are in tons per day.
Numbers in parentheses are the production run numbers.

recycling. Second, the new newsprint plant is sensitive to higher levels of both ambient water and air quality standards, i.e., activity level reduces to zero for all combinations above E-E, except for E-water combined with T-air, because the costs of residuals management at the newsprint plant using used newspapers increase rapidly at higher environmental quality standards.

Linkages

One of the original objectives of applying the REQM modelling framework to the Lower Delaware Valley region was to investigate the linkages among gaseous, liquid, solid, and energy residuals, and among the various environmental media, in a real-world situation. The evidence of linkages among the forms of residuals and among the three environmental media may be obtained from a comparison of: objective function values; magnitudes of flows in the region, such as fuels, residuals discharges, residuals transport (mixed solid residuals and sludge barging), and used paper; and/or capacities of structures such as municipal incinerators, regional sewage treatment plants, and add-on residuals modification devices (e.g., electrostatic precipitators).

Suggested evidence of linkages is presented in figure 3-8, A and B. In A, objective function values for the region are shown for varying combinations of ambient air and water quality standards, with high quality landfills. With existing ambient air quality standards (i.e., the "0" level), it costs the region about $76,000 per day to go from existing conditions to the E-level water quality standards. When E-level air quality standards are imposed, the costs of improving water quality to the E-level are about $100,000 per day, a difference of about $24,000. Had there been no linkages

Figure 3-8. Evidence of Linkages Among Forms of Residuals: Lower Delaware Valley Model with High Quality Landfills

-- A --

Air Quality Standards

Water Quality Standards	Total Regional Costs, 10^6 \$/day: O	Total Regional Costs, 10^6 \$/day: E	Incremental Regional Costs, 10^6 \$/day: O	Incremental Regional Costs, 10^6 \$/day: E
O	0.034 (#25)	0.26 (#12)		
			0.076	0.10
E	0.11 (#10)	0.36 (#4)		
			0.03	0.06
T	0.14 (#11)	0.42 (#3)		

-- B --

Air Quality Standards

Water Quality Standards	Total MPSX 1 Costs, 10^6 \$/day: O	Total MPSX 1 Costs, 10^6 \$/day: E	Incremental MPSX 1 Costs, 10^6 \$/day: O	Incremental MPSX 1 Costs, 10^6 \$/day: E
O	0.0 (#25)	0.11 (#12)		
			0.020	0.04
E	0.020 (#10)	0.15 (#4)		
			0.004	0.02
T	0.024 (#11)	0.17 (#3)		

Note: Numbers in parentheses are the production run numbers.

at all between air and water, the increased costs would have been the same for both cases. In going from E-level water quality to the T-level, the difference is larger: about $30,000 per day when air quality is not considered, compared to about $60,000 per day when the E-level air quality standard is imposed, a difference of about $30,000.

In B, evidence is based on the costs from the first module (MPSX 1), containing the petroleum refineries, the steel mills, and the electric power plants. When no air quality standards are imposed, the cost to move from the O-level to E-level water quality standards is about $20,000 per day. When E-level air quality standards are imposed simultaneously, the incremental costs to achieve E-level water quality standards are about $40,000 per day. In moving from E-level to T-level water quality standards when air quality standards are not imposed, the incremental cost is about $4,000 per day. When E-level air quality standards are imposed simultaneously with T-level water quality standards, the incremental cost is about $20,000 per day. Again, the linkages between air quality and water quality in the region are clearly indicated.

The final evidence of linkages presented here is the interrelationship between different levels of landfill quality and ambient air and water quality standards. Data are presented for two cases in figure 3-9: one for E-level air and water quality standards; one for T-level air and water quality standards. If there were no linkages between landfill quality and air and water quality, the difference between total costs (column 2) and the difference between solid residuals management costs (column 3) would have been the same. That they are substantially different indicates that linkages exist. Thus, substantial

Figure 3-9. Relationship between Solid Residuals Management and Air and Water Quality Management: Lower Delaware Valley Model

Easy (E) Air and Water Quality Standards		
	Total costs to region, 10^6 \$/day	Solid residuals management costs (MPSX 5), 10^6 \$/day
High quality landfill (#4)	0.36	0.060
Medium quality landfill (#8)	0.31	0.038
Difference	0.05	0.022

Tight (T) Air and Water Quality Standards		
	Total costs to region, 10^6 \$/day	Solid residuals management costs (MPSX 5), 10^6 \$/day
High quality landfill (#1)	1.20	0.079
Medium quality landfill (#5)	1.16	0.055
Difference	0.04	0.024

Note: Numbers in parentheses identify the production runs.

costs are imposed on other sectors when higher quality landfills are required, the linkages to these additional costs being through the air and water quality of the region.

Distributional Impacts

With respect to the distribution of AEQ, table 3-9 shows the distribution of SO_2 concentrations among the fifty-seven political jurisdictions for runs 1 and 3, the T-level air quality standards and the E-level air quality standards, respectively. The maximum concentration for both runs is found in political jurisdiction 16 (South Philadelphia). Of the highest concentrations for run 1, eight are located within Philadelphia (jurisdictions 14, 15, 16, 17, 18, 19, 23, and 27), one in Delaware County (jurisdiction 9), and one in Camden, New Jersey (jurisdiction 50). Although not shown in this table, the TSP distribution for run 1 is similar to that for SO_2, in that of the ten highest concentrations, nine are located within Philadelphia (jurisdictions 14, 15, 16, 19, 22, 23, 24, 26, 27) and one in Camden, New Jersey (jurisdiction 50). Unlike SO_2, however, the maximum concentration of TSP for runs 1 and 3 is found in jurisdiction 15 (in Center City Philadelphia). These results demonstrate that, for virtually all reasonable REQM strategies, the poorest air quality throughout the region will still be found in the Philadelphia-Camden area.

The distributions of dissolved oxygen (DO) concentrations throughout the 22-reach estuary for runs 1, 2, 3, 13, 15, and 16 are shown in table 3-10. When E-level water quality standards are imposed (run 2), the eight poorest water quality reaches (which actually are of fairly high quality), are all located between reaches 11 and 19 (see figure 3-2) from the confluence with the Schuylkill River to below Wilmington. Using DO as an indicator of water

Table 3-9. Distribution of Sulphur Dioxide Concentration by Political Jurisdiction for Selected Production Runs

Political Jurisdiction	Sulphur dioxide (μgms/m^3)		Political Jurisdiction	Sulphur dioxide (μgms/m^3)	
	Run 1	Run 3		Run 1	Run 3
1	17	54	30	23	66
2	15	49	31	16	46
3	6	14	32	13	44
4	12	39	33	12	40
5	8	22	34	11	35
6	5	19	35	11	32
7	5	12	36	6	20
8	22	53	37	12	37
9	32*	68	38	9	27
10	25	58	39	4	12
11	28	64	40	11	46
12	15	42	41	9	34
13	11	32	42	7	24
14	56*	96*	43	4	11
15	42*	98*	44	13	40
16	83*#	120*#	45	10	34
17	64*	100*	46	8	20
18	35*	70	47	12	49
19	50*	91*	48	11	40
20	24	57	49	8	25
21	18	48	50	40*	87*
22	29	74*	51	20	64
23	38*	89*	52	27	71
24	29	103*	53	16	43
25	19	66	54	6	18
26	25	70	55	22	51
27	34*	94*	56	9	26
28	25	65	57	8	26
29	17	49			

*Indicates the 10 highest concentrations of sulphur dioxide.
#Indicates the maximum concentration of sulphur dioxide.

Table 3-10. Distribution of Water Quality by Estuary Reach for Selected Production Runs--Dissolved Oxygen

Delaware Estuary Reach[a]	Dissolved oxygen (mg/ℓ)					
	Run 1	Run 2	Run 3	Run 13	Run 15	Run 16
1	8.6	8.6	8.6	8.6	8.6	8.6
2	7.7	7.7	8.0	7.9	7.9	7.9
3	6.9	6.6	7.2	7.0	6.9	7.2
4	6.3	6.0	6.8	6.2	6.1	6.5
5	5.9*	5.7	6.4	5.8*	5.4*	6.0
6	6.0	5.9	6.5	6.0	5.4*	6.0
7	5.9*	5.9	6.3	5.8*	5.3*	6.0
8	5.9	5.8	6.3	6.2	5.7	5.8*
9	6.4	6.1	6.7	6.8	6.3	6.2
10	6.8	5.3	7.0	7.8	7.5	7.1
11	5.3*	3.6*	5.1*	5.7*	5.2*	5.1*
12	6.1	3.7*	5.7*	6.3	6.0	6.1
13	5.7*	3.6*	5.4*	5.7*	5.3*	5.0*
14	5.7*	4.0*	5.5*	5.7*	5.3*	5.1*
15	6.1	4.5*	6.0*	6.0	5.5*	5.8*
16	6.4	5.2	6.3	6.0*	5.6	6.4
17	5.0*#	3.0*#	5.0*#	4.9*#	5.0*#	5.0*#
18	5.1*	3.4*	5.2*	5.4*	5.5	5.1*
19	5.7*	4.8*	5.8*	6.0	6.1	5.8*
20	6.1	5.8	6.1	6.1	6.2	6.1
21	6.2	6.2	6.3	6.3	6.3	6.3
22	6.6	6.6	6.6	6.7	6.7	6.6

[a]For locations of reaches, see figure 3-2, page 59.

*Indicates the 8 lowest concentrations of dissolved oxygen.

#Indicates the minimum concentration of dissolved oxygen

quality, estuary water quality upstream from the confluence with the Schuylkill River is better than downstream. The poorest quality reach is 17, near Wilmington, Delaware.

For the T-level water quality standard (DO$\geq$ 5.0 mg/ℓ), the distribution of the poorer quality reaches changes (runs 1, 3, 13, 15, and 16). Now the 8 poorer water quality reaches are distributed between reaches 5 and 19, from above Philadelphia to below Wilmington, but the distributions differ depending on whether or not regional sewage treatment plants and/or in-stream aeration are employed. As before, the poorest water quality reach is still 17 near Wilmington. As with the E-level standards, the poorest quality water is not found near Philadelphia.

With respect to the distribution of costs, table 3-11 shows for each of the 57 political jurisdictions:

1. the percentage increases in home heating costs under:

   a. T-level air quality standards (run 1); and
   b. E-level air quality standards (run 3);

2. the percentage increases in household electricity costs under the same two situations as for home heating costs;

3. the percentage increases in municipal solid residuals management costs for:

   a. high quality landfills with E-level air quality standards (run 3); and

   b. medium quality landfills with E-level air quality standards (run 7); and

4. the increase in costs of sewage disposal per sewered household under:

   a. T-level water quality standards (run 1), and
   b. E-level water quality standards (run 2).

For the T-level air quality standards (run 1), the maximum increase in household electricity costs, about 16 percent, is found in New Castle County, Delaware (jurisdictions 1-4). The second highest increase, about 14 percent,

Table 3-11. Distribution of Increased Consumer Costs by Political Jurisdiction for Selected Production Runs

Political jurisdiction	Home heating (percent)		Household electricity (percent)		Municipal solid residuals mgmt. (percent)		Municipal sewage ($/household/year)	
	Run 1[a]	Run 3	Run 1	Run 3	Run 3	Run 7	Run 1	Run 2
1	25.1	1.8	15.8**	2.4	19.2	7.7	0.5*	0.1
2	30.5	0.7	15.8**	2.4	19.4	7.8	43.5	8.0
3	29.0	0*	15.8**	2.4	19.3	7.7	5.9	1.1
4	23.0	0*	15.8**	2.4	19.4	7.8	1.0	0.2
5	33.9	0.6	13.6	4.5**	19.1	7.6	n.a.	n.a.
6	35.5	0*	13.6	4.5	19.1	7.7	n.a.	n.a.
7	16.8	0*	13.6	4.5	19.4	7.8	n.a.	n.a.
8	26.1	9.6	13.6	4.5	10.8	7.7	3.3	2.5
9	23.3	7.2	13.6	4.5	17.9	7.7	2.1	1.0
10	24.5	3.2	13.6	4.5	19.3	7.7	3.5	1.9
11	28.3	0.9	13.6	4.5	10.9	7.7	16.2	8.0
12	27.8	3.4	13.6	4.5	18.8	7.7	3.9	2.1
13	26.2	1.9	13.6	4.5	21.0	7.7	2.5	1.7
14	19.7	6.1	13.6	4.5	8.7	7.7	21.4	16.1
15	41.0	25.8	13.6	4.5	7.6*	7.6	16.2	12.2
16	31.4	18.5	13.6	4.5	17.3	7.7	15.4	10.6
17	16.4	7.4	13.6	4.5	8.1	7.7	19.3	13.1
18	21.7	8.2	13.6	4.5	17.5	7.7	18.8	12.8
19	50.4	28.5	13.6	4.5	20.1	7.6	14.4	9.8
20	22.2	6.3	13.6	4.5	20.4	7.7	16.8	11.4
21	25.6	6.3	13.6	4.5	8.6	7.6	16.6	11.6
22	31.7	15.2	13.6	4.5	20.4**	7.7	19.3	14.0
23	55.4**	30.0**	13.6	4.5	11.2	7.5	14.8	10.6
24	14.1	0.2	13.6	4.5	17.6	6.6*	8.0	7.1
25	10.4	0.5	13.6	4.5	17.8	6.6*	16.0	14.2
26	12.6	0.9	13.6	4.5	23.1	8.2**	15.0	13.4

continued ...

Table 3-11. (Continued)

Political jurisdiction	Home heating (percent)		Household electricity (percent)		Municipal solid residuals mgmt. (percent)		Municipal sewage ($/household/year)	
	Run 1[a]	Run 3	Run 1	Run 3	Run 3	Run 7	Run 1	Run 2
27	25.5	4.8	13.6	4.5	20.5	7.7	22.1	16.7
28	29.8	5.1	13.6	4.5	20.4	7.7	15.6	13.3
29	33.9	1.6	13.6	4.5	10.7	7.6	17.9	14.1
30	16.5	2.7	13.6	4.5	20.6	7.7	15.6	13.9
31	17.2	2.7	13.6	4.5	20.3	7.7	15.7	13.6
32	9.9*	0.6	13.6	4.5	20.2	7.6	15.8	14.0
33	31.4	1.1	13.6	4.5	20.4	7.7	19.7	17.6
34	35.1	1.9	13.6	4.5	18.7	7.5	7.3	4.9
35	26.9	2.8	13.6	4.5	19.1	7.7	2.3	1.5
36	31.2	0.0	13.6	4.5	19.2	7.7	n.a.	n.a.
37	26.7	0.7	13.6	4.5	19.0	7.6	6.1	5.4
38	26.9	0.6	13.6	4.5	19.0	7.6	1.4	1.2
39	36.7	0.0*	n.a.	n.a.	19.2	7.7	n.a.	n.a.
40	36.7	0.2	13.6	4.5	19.1	7.6	8.3	1.3
41	31.3	0.0*	13.6	4.5	19.2	7.7	15.9	0.0*
42	29.6	0.3	13.6	4.5	19.3	7.7	1.8	1.6
43	33.1	0.0*	6.8	2.2	19.3	7.7	n.a.	n.a.
44	34.9	0.0*	2.9	0.4	19.4	7.7	48.2**	1.2
45	24.2	0.0*	2.9	0.4	19.2	7.7	n.a.	n.a.
46	25.0	0.0*	2.9	0.4	19.2	7.7	n.a.	n.a.
47	25.1	0.8	2.9	0.4	19.3	7.7	12.7	3.0
48	17.4	0.0*	2.9	0.4	19.5	7.8	11.5	0.0*
49	19.7	0.0*	1.4*	0.2*	19.4	7.8	2.9	0.0*
50	31.7	9.2	2.9	0.4	19.5	7.8	47.8	36.9**
51	20.6	4.1	2.9	0.4	19.3	7.7	20.7	13.6
52	29.5	6.0	2.9	0.4	19.4	7.8	4.4	3.2
53	22.3	2.8	2.9	0.4	19.4	7.8	n.a.	n.a.
54	25.6	0.7	10.6	1.3	19.3	7.7	n.a.	n.a.

continued ...

Table 3.11 (Continued)

Political jurisdiction	Home heating (percent)		Household electricity (percent)		Municipal solid residuals mgmt. (percent)		Municipal sewage ($/household/year)	
	Run 1[a]	Run 3	Run 1	Run 3	Run 3	Run 7	Run 1	Run 2
55	25.9	5.0	2.9	0.4	19.5	7.8	6.6	4.5
56	19.2	0.2	10.6	1.3	19.5	7.8	n.a.	n.a.
57	28.2	0.0	10.6	1.3	19.6	7.9	0.5	0.1

[a] Cost increases for home heating are all at the technological upper limits except for jurisdictions 1, 2, 7, 12, 35, 49, and 56. There is empirical evidence to suggest that at the "optimum" for this run, all cost increases would be at their upper limits. Cost increases at the upper limits represent a situation where there is total conversion within the region to natural gas.

* Minimum increases

** Maximum increases

n.a. = not applicable

is found in Chester, Delaware, Philadelphia, Montgomery, and Bucks counties in Pennsylvania (jurisdictions 5-42). For the E-level air quality standards, the situation reverses, and the Pennsylvania counties, including Philadelphia, are associated with the maximum increase in costs (4.5 percent). Hence, as with water quality, the maximum cost increases are not always associated with Philadelphia. Furthermore, in Camden, New Jersey, where the air quality is almost as poor as in Philadelphia, the increased household electricity costs for the T-level air quality standards amount to only 2.9 percent.

For municipal sewage disposal, among the ten jurisdictions exhibiting the highest average increased costs ($ per household per year) are Wilmington (jurisdiction 2), Trenton (jurisdiction 44), and Camden (jurisdiction 50).[8] In fact, one of these three cities is always in the number one position--the maximum increase in costs of sewage disposal. The remaining seven highest cost jurisdictions are, however, located within the City of Philadelphia.

There is relatively little variation among the jurisdictions in the increased costs of municipal solid residuals management. This is because, for all runs, all jurisdictions were required to maintain the same quality landfills, i.e., high quality, medium quality, or low quality. Where a significant variation did occur, it was associated with the high landfill quality runs, such as run 3. Within this group of runs, the differences in increased costs among jurisdictions for the E-level air quality standards were substantially greater than those associated with the T-level standards. This is because for the E-level air quality standards seven municipal incinerators (out of a total of twenty-three), all available to Philadephia, were used for solid residuals management. The result of this is that the solid residuals disposal costs of the six jurisdictions within Philadelphia that use these incinerators are roughly 50 percent of those jurisdictions that do not have this option. However, when T-level air quality standards are imposed, some of the incinerators may shut down, resulting in more uniform cost increases throughout the region.

Conclusions

The development of the regional REQM model and its application to the Lower Delaware Valley region has resulted in two sets of conclusions. The first set relates to REQM for the Lower Delaware Valley region; the second relates to methodological issues in analysis for REQM.

Conclusions: REQM for the Lower Delaware Valley Region

1. Air quality over the Lower Delaware Valley region appears to be more of a problem than Delaware estuary water quality for at least two reasons. First, in some areas of Philadelphia, it appears that it may be technologically difficult to achieve, and then to maintain, the federal primary annual average standards for sulphur dioxide and suspended particulates. Second, regional costs to achieve the federal primary air quality standards would be 6 to 8 times the regional costs to achieve water quality standards in the estuary, between $300 and $400 million per year versus about $50 million per year, respectively.

2. Estuary water quality standards can be met in all cases examined by using various combinations of: on-site residuals discharge reduction; regional sewage treatment; and in-stream aeration. The regional alternatives appear to allow the region as a whole to meet the same ambient water quality standards more cheaply than it can by relying on measures at individual sites, but not all dischargers benefit to the same degree. Indeed, some dischargers who are not serviced by the regional facilities actually pay

more under management alternatives with regional facilities than when management alternatives are restricted to local sewage treatment.

3. Under the conditions reflected in the model, roughly those of 1970, it appears that the rail haul alternative for disposing of solid residuals is too expensive. However, as landfill sites become scarce (and hence more expensive), this alternative will certainly become more attractive, as will the alternative of increased recycling.

4. Again, under 1970 conditions, it appears that new municipal incinerators are not economically feasible alternatives for solid residuals management, especially with the stringent standards which would be imposed on liquid and gaseous discharges.

5. Given the goal to improve air and water quality in the region, and the inevitable increases in landfill costs, increased use of used paper products in the Lower Delaware Valley appears to be a promising management option.

6. Based on the set of analyses presented, there is a variety of combinations of physical measures for meeting the regional AEQ standards that do not differ very much in total costs to the region. However, the distributions of levels of residuals discharge reduction, of residuals discharges, and of costs among dischargers, all vary substantially from combination to combination.

Conclusions: Methodological Issues

1. Intermedia Linkages and Tradeoffs

Evidence of the importance of linkages and tradeoffs among the ambient qualities of the three environmental media is provided by the results of the analyses. The examples presented demonstrate the kinds and extent of the interdependencies among residuals forms and among environmental media, but the question remains: is it really worth the effort to include these linkages explicitly in the analyses? One part of the question is, how difficult and costly is it to include these linkages in the analysis of alternative regional REQM strategies? The other part of this question is, how costly might it be to the region if these linkages were not considered explicitly?

With respect to the first part, the Lower Delaware Valley study indicated that the marginal costs of including the air-water-solids linkages were modest, to the extent that the costs can be separated at all. There were no problems in obtaining the linkage data themselves, such as flyash generation rates in sludge incineration, or suspended solids loads created in wet stack-gas scrubbers. The one drawback of the linked models is their size, which has implications for the costs of analysis. But all things considered, the benefit-cost ratio for linked models appears very favorable to further development and application, especially where AEQ standards are involved.

The second part of the question is considerably more difficult to answer. No direct attempt was made in the Lower Delaware Valley study to obtain an answer.[9]

2. Nonlinear Aquatic Ecosystem Models

The potential value of using aquatic ecosystem models in an REQM context is very great. Compared with classical linear dissolved oxygen models, these newer models are potentially capable of providing additional information on the resulting state of the natural environment, such as algal densities and fish biomass. Also, they are alleged to provide more accurate predictions of dissolved oxygen over wider ranges of river flows and residuals discharges than the linear dissolved oxygen models. But there are substantial costs involved in incorporating them in a regional management model. The inclusion of nonlinear ecosystem models within an optimization framework creates substantial computational problems and expense. In addition, at the current state-of-the art, these models do not provide accurate predictions at trophic levels above algae.

Whether or not it is worth the problems and expense of incorporating aquatic ecosystem models depends on the region being studied and the question being addressed. For the larger and more complex regions, a reasonable compromise at this time might be to use a linear dissolved oxygen water quality model within the optimization framework. This would at least provide a relationship in the model between organic discharges and one indicator of ambient water quality. Then, using the "optimal" set of organic discharges from the regional model as input to an ecosystem simulation model, the implications for other water quality indicators, such as algal densities, could be investigated.

Another possibility would be to employ a linear phytoplankton model in the regional analysis. These models do exist, but they are not in as widespread use today as the nonlinear variety. In addition, certain restrictions must be placed upon their use in the analysis because of the assumptions made in their development.

The feasibility of using nonlinear aquatic ecosystem models in regional REQM analyses would be enhanced by the development of efficient, large-scale, nonlinear programming algorithms that could deal with resource management problems of the type described here.

3. Distribution Information

The capability of providing distributional information on both REQM costs and AEQ is one of the most important features of the regional REQM model presented here. In most regions, the distribution of costs and of AEQ will be a more important issue than regional economic efficiency. Unfortunately, the provision of this additional information was not without computational problems.

Relatively little difficulty was encountered in constraining levels of AEQ and generating information on the implied costs by political jurisdiction in the region. Of course, it would have been still easier, and certainly less expensive, merely to provide information on the implied costs and implied levels of AEQ for various alternative REQM strategies, however selected. But the real difficulty arose when the attempt was made to constrain both the levels of AEQ and the distribution of costs simultaneously. There are two primary reasons for this difficulty. The first involves infeasible solutions. This will not be discussed here.[10]

The second computational difficulty is associated with the fact that most nonlinear programming algorithms become less and less efficient as the optimum is reached. When the regional efficiency criterion is employed, it makes sense to stop these algorithms short of an optimum because only modest cost savings (as a percentage of total regional costs) are involved. However, significant shifts in the distribution of costs continue to occur with virtually every step in the approach to the optimum. Thus, stopping short of the optimum makes it difficult, if not impossible, to determine when comparable results of runs for different sets of conditions have been obtained. In addition, stopping short of the optimum makes it difficult to explore the tradeoffs among the distributions of costs because these tradeoffs occur at the flattest portion of the regional cost response surface, near the optimum. Therefore, for analysis of regional REQM where specific information on distribution of both costs and of AEQ is desired, resort must be made to linear programming techniques where optimum solutions can be obtained relatively efficiently. This means the elimination of nonlinear models of both the natural world and of residuals generation and discharge as components of regional REQM optimization models.[11]

Footnotes

[1] The complete discussion of the Lower Delware Valley study and a list of related publications can be found in, Walter O. Spofford, Jr., Clifford S. Russell, and Robert A. Kelly, Environmental Quality Management/An Application to the Lower Delaware Valley, Research Paper R-1 (Washington, D.C.: Resources for the Future, 1976).

[2] The definition of "urbanized" used by the U.S. Bureau of the Census is long and complex, but basically it amounts to counting as urbanized incorporated places of more than 2,500 inhabitants or greater than 1,000 people per square mile density.

[3] In the modelling application, 1970 was used as the base period, and the structure reflects the situation in that year whenever possible. In some cases, such as for sewage treatment plants, in which government surveys were the data source, the base year differs from 1970.

[4] It should be noted that this use of uniform wind rose and stability conditions for the region, in combination with the available source inventory data, produced distributions of ambient air quality similar to the measured annual averages.

[5] At 1975 computer rates, it cost $1,220 for a 30-iteration run, or roughly $41 per iteration, operating on an IBM 370 model 165 and using 300K bytes of internal core storage. Per iteration, it takes, on average, 2.80 minutes of CPU time, 5.94 minutes of I-O time, has 13,400 read-write instructions, and prints 1,236 lines of output (minimum).

[6] The one exception to the total discharges not changing much beyond the 30th iteration is the trade-off between in-stream aerators and BOD_5 discharges (and perhaps other discharges).

[7] This is particularly true, because it was assumed, at the time the model inputs were developed, that natural gas was available to all users at a relatively low price.

[8] The costs of sewage disposal have been allocated in the model on the basis of both flow and BOD_5 load. The BOD_5 concentrations of sewage have been assumed using different loading factors for urban, suburban, and rural areas. The average increased costs to jurisdictions for sewage disposal is computed on the basis of total sewage disposal costs divided by total housing units for each jurisdiction. Thus, all other factors the same, jurisdictions that are less than 100 percent sewered will register smaller average increases than those that are 100 percent sewered. This accounts for the apparent discrepancy in increased costs for jurisdictions served by the same sewage treatment plant.

[9] However, there is some discussion of the issue in Spofford, et al, op. cit.

[10] See Spofford, et al, op. cit.

[11] This statement assumes that the nonlinearities are such that they cannot be converted into piecewise linear segments for inclusion in a linear programming model.

Chapter 4

A COORDINATED SET OF ECONOMIC, HYDRO-SALINITY AND AIR QUALITY MODELS OF THE UPPER COLORADO RIVER BASIN WITH APPLICATIONS TO CURRENT PROBLEMS

Charles W. Howe*

Physical and Economic Description of the Region

The Upper Colorado River Basin has an area of approximately 102,000 square miles, located in southwestern Wyoming, western Colorado, eastern Utah, northwestern New Mexico, and northeastern Arizona. It extends from latitude 35° 34' north to 43° 27' north, a distance of about 540 miles; and from longitude 105° 38' west to 112° 19' west, a distance of about 350 miles. Figure 4-1 shows the entire Colorado River Basin, of which the Green, Upper Main Stem, and San Juan River Basin comprise what is called the Upper Basin. The basin is bounded on the east by the main Rocky Mountain chain, on the west by several high plateaus and the Wyoming and Wasatch ranges.

The basin is sparsely populated with about 346,000 persons (1970 census) averaging 3.4 persons per square mile, compared with a national average density of 57.4. The low density is primarily due to the mountainous terrain and the arid to semiarid climate of much of the remainder of the region.

The northern part of the basin is largely mountainous plateau at 5,000 to 8,000 feet elevation, including broad rolling valleys and high intersecting

*This paper covers work undertaken jointly by Dr. Jan Kreider, Professor Bernard Udis, and the author, all of the University of Colorado, under the sponsorship of the Economic Development Administration, U.S. Department of Commerce, Grant OER 351-G-71-8(99-7-13215).

Figure 4-1. Major Subbasins of the Colorado River Basin

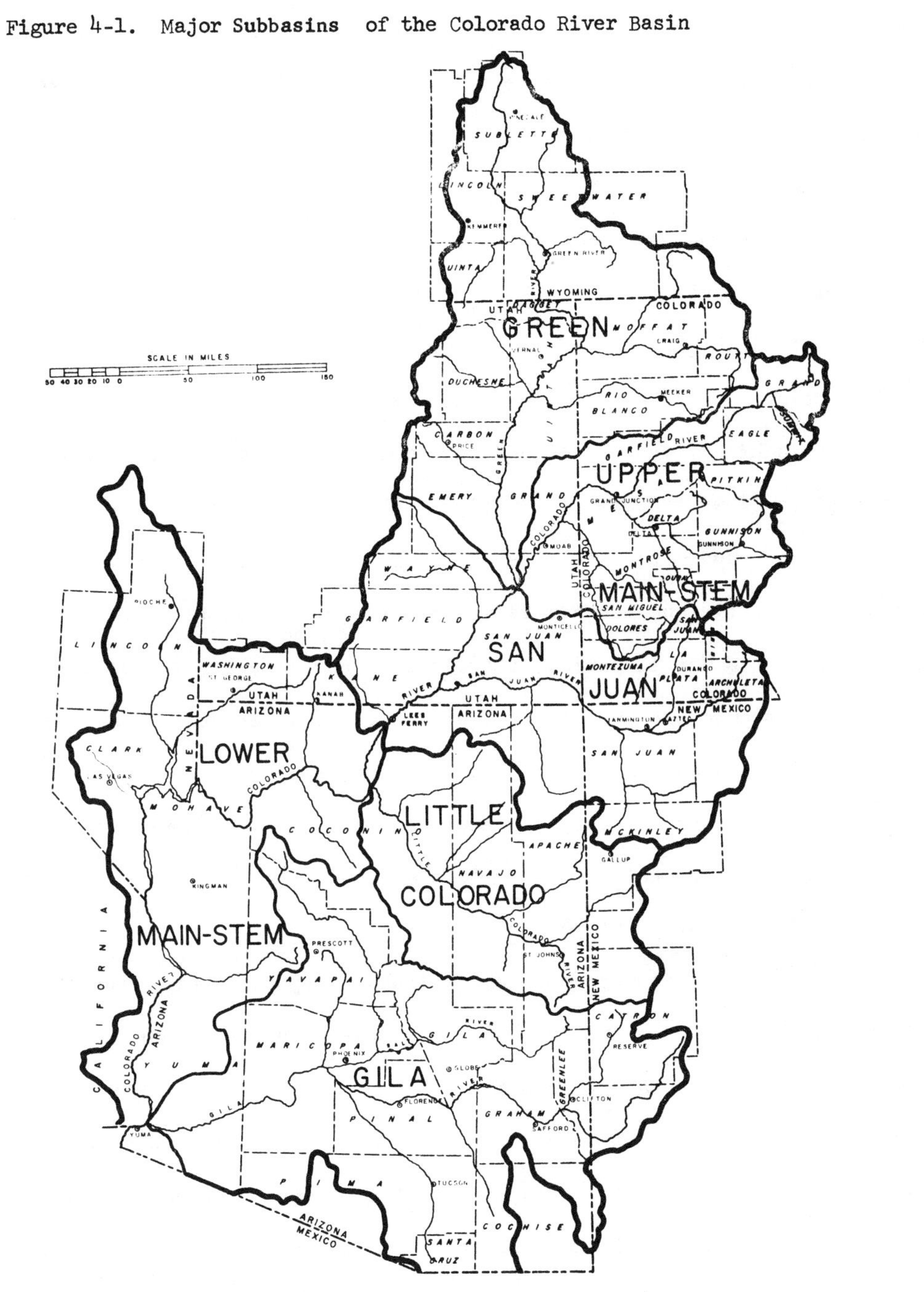

Source: Colorado River Basin Water Quality Control Project, Federal Water Pollution Control Administration, U.S. Department of the Interior, San Francisco, Calif.

mountain ranges with many peaks above 14,000 feet. The southern part of the basin has broad alluvial valleys and rolling plateaus broken by mountain ranges. The main river and major tributaries tend to flow through deep canyons.

The climate ranges from continuous snow cover and heavy precipitation on the western slopes of the Rocky Mountains to desert conditions in the south. Most of the moisture falls as winter snow, with spring and summer rainfall being localized infrequent storm activity. Water supply in the basin thus is highly dependent upon construction of dams, ground water, and transfers of water among the various subbasins.

The major economic activities of the basin traditionally have been agriculture, cattle and sheep raising, and the mining of metallic minerals. In the post-World War II period, recreational use of all parts of the basin has expanded tremendously, ranging from exclusive international skiing resorts like Aspen and Vail to mountain hiking, fishing, open river and reservoir boating, and desert area exploration. This broad recreational use is facilitated by the extensive federal government land holdings which constitute 70 percent of the basin's area.

Most recently, the huge coal and oil shale deposits which underlie much of the basin have attracted intensive development pressure. The federal government is intent upon expansion of the use of the energy resources. The state governments of the area are less certain about the desirability of large-scale strip mining of coal with associated power plants and the development of oil shale refining. Environmental problems could be critical, water requirements would be large, and interference with the swift growth of recreation and tourism is imminent.

The availability of water for further economic expansion of the basin is a major issue. At the time of the original agreement concerning the division of the waters of the Colorado River (The Colorado River Compact ratified in 1929), the average annual water available was thought to be 15 million acre-feet per year (maf/yr); 7.5 maf/yr were allotted to the Lower Basin. Since that time, long-term average flows have been decreasing, until it is now felt that the average annual availability may be as low as 12.5 maf. In addition, a U.S.-Mexican treaty of 1944 calls for a guaranteed delivery of 1.5 maf/yr to Mexico, and it is not clear how this obligation is to be divided between Upper and Lower Basins, although it has been declared a "national obligation" by recent legislation. It is probable that about 5.0 million acre-feet per year is legally available for consumptive use in the Upper Basin, with 2.5 to 3.0 maf currently being consumptively used.

The fact that the Upper Basin is not fully consuming the water available does not mean that the water is going unused. Table 4-1 shows the current annual actual and allotted rates of consumptive use in the Lower Basin. Adding this total of 10.8 maf/yr to the 2.5 to 3.0 maf/yr of Upper Basin consumptive use makes clear that the river's flow is now fully utilized, and indeed, in a typical year no flow whatsoever reaches the river's original terminus, the Gulf of California.

As further development takes place in the Upper Basin, some of the current uses of water will have to be foregone. Southern California is a prime candidate to give up water use, since that state has a legal right to only 4.4 maf/yr while currently using 5.2 maf/yr. It has been calculated that the opportunity cost of marginal withdrawals of water from

Table 4-1. Current* Annual Actual and Allotted Consumptive Rates of Use of Colorado River Water in the Lower Basin (maf/yr)

California (current use)	5.2
Arizona (allotment)	2.8
Nevada (allotment)	0.3
Mexico (treaty)	1.5
Evaporative losses	1.0
	10.8

*current = approximately 1970

agriculture in Southern California ranges from $10 to $100 per acre-foot, depending upon the availability of substitute commodities as inputs into the agricultural and food processing industries and the mobility of resources out of agriculture.[1] Additional water demands in the Upper Basin could also be met through reductions in current water uses in the Upper Basin itself, a strategy the value of which is investigated later in this paper.

Certainly Upper Basin demands will grow. Large expansions of thermal-electric power are planned, based upon the extensive coal deposits, and even larger expansions will be necessitated if shale oil development proceeds. The U.S. government continues to mention a shale oil output capacity of 1.5×10^6 bbls/day, a capacity which would require directly about 210,000 af/yr, plus the very sizeable indirect consumptive uses related to the associated electric energy generation and population. Coal gasification plants already under construction in the San Juan subbasin will consume over 200,000 af/yr.

It is very clear, therefore, that water is a scarce commodity with an increasing opportunity cost to all emerging uses. Patterns of regional development will be constrained by water availability, so it is important to investigate the economic, environmental, and equity implications of alternative patterns of growth and what the trade-offs are.

Accompanying the growth of water use in the Upper Basin, there has been a deterioration of water quality in the form of a rising trend of total dissolved solids (TDS).[2] Current[3] TDS levels at Imperial Dam in the Lower Basin average 850 ppm and are predicted to rise to 1100 ppm by the year 2000, and perhaps higher if the consumptive uses and likely TDS additions of the shale oil industry occur. Current levels are 50 percent attributable to natural point and diffuse sources, 40 percent to agricultural return flows, and 10 percent to municipal-industrial uses and out of basin transfers.

A major program of salinity reduction is being planned by the Bureau of Reclamation. This program involves control of several major natural point sources (salt springs and large natural salt deposits along river beds) and increased efficiency of water use in the agricultural sector in the Upper Basin. There are agricultural areas which are shallowly underlain by salt deposits. When these lands are irrigated, percolating water dissolves large quantities of salt which then is added to the stream by the return flow. Additions of dissolved salts of as much as 8 tons per year per acre irrigated are found in the Grand Valley of western Colorado.

The United States and Mexico have had extensive conflicts over the quality of the Colorado River as it passes into Mexico. While an upward trend in TDS had been observed for years, in 1960 the TDS content at the Mexican border leaped to 6000 or more ppm as a result of the Wellton-Mohawk Irrigation Project. Agricultural production in the Mexicali Valley was severly damaged, and farming remains impaired even at the lower levels (1500 ppm) which have been achieved at substantial cost. The United States has agreed to improve the water quality for Mexico even more, one possibility being the construction of a large desalting plant to treat the return flows from the Wellton-Mohawk Project.

Thus it is quite clear that the external effects of any developments which increase TDS in the Colorado River are no longer negligible and must be taken into account in any socially responsible and economically efficient regional planning.

In addition to the water quantity and quality dimensions of regional growth, air quality has become an important consideration. Much of the U.S. Southwest has historically been characterized by clean air, great distances, and grand vistas of the spectacular landscape. The clean air of the mountain valleys was a valued amenity. In the late 1960's, very large thermal electric plants began appearing in the high plateau country of the so-called Four Corners Area.[4] These plants were of 350 to 500 million watts (mw) capacity, designed to produce electric energy for Los Angeles, Phoenix, and other urban centers. The early plants had no equipment to reduce discharges of gaseous residuals, so that it was but a short time until vast plumes of particulate matter and sulphur oxides could be traced hundreds of miles from the plants. All plants have been required now to utilize residuals discharge reduction equipment, but with a total of 20,085 mw capacity now installed, a region once characterized by clarity of its air now suffers from a pall of particulate haze and environmental damage from the gaseous residuals.

Further north, localized poor air quality conditions in the mountain valleys seriously impair the quality of life. The famous resorts of Vail and Aspen often suffer inversions which require bans on commercial fuel use and private fireplaces. While there are few industrial activities discharging gaseous residuals at present, proposed shale oil and related thermal electric stations will add large loads to the air, and the siting of these facilities will be crucial from an air quality management viewpoint.

Observations on the Institutional Setting of the Region

The Upper Colorado River Basin represents part of five states: Wyoming; Colorado; Utah; New Mexico; and Arizona. Each state has its own system of water law governing the establishment of legal title to surface and ground waters, although all laws are "appropriation doctrine", which doctrine permits claims to water by individual parties in priority according to date of first use.

While the waters of the entire river were distributed between Upper and Lower Basins by the 1929 Colorado River Treaty, an Upper Basin Compact of 1954 has allocated the Upper Basin waters among the states. A United States Supreme Court decree in 1964, resulting from a long-standing lawsuit among the Lower Basin states, allocated the Lower Basin's 7.5 maf/yr among California (4.4 maf), Arizona (2.8 maf), and Nevada (0.3 maf). Each state seeks to protect its share of the water. In addition, the federal government has asserted some rather nebulous but potentially sizeable claims because of its vast land holdings.

Thus there is no river basin agency which has management responsibility for the entire basin, and water has been legally allotted to individual states. This institutional setting has the following effects:

1. States which have claims to water in excess of their current uses are eager to put their water to use regardless of efficiency considerations, for fear that some change of law may deprive them of their unused water.

2. Water cannot be reallocated among states or between Upper and Lower Basins without very substantial changes in existing laws and compacts.

3. States are not concerned with the downstream effects of their actions, either in terms of reducing existing downstream water uses or the damages from increased salinity.

The same failures to acknowledge the opportunity costs of water and external salinity damages hold within a given state among water users. For reasons too complex to discuss here, the water rights established by different users are difficult to buy and sell if any change in type and location of use is involved. Markets for water rights are thus quite imperfect, transactions costs are high, and opportunity costs are very poorly reflected. Individual users (90 percent agricultural) have little incentive to maintain high physical efficiency in use. In like manner, there are no penalties for increasing the TDS levels of streams through return flows, so there is no private incentive to avoid those irrigated agricultural lands which add so much to the salinity loads.

With respect to air quality management, while federal and state ambient air quality standards have been set, reliance on state enforcement has resulted in a slowness to approach the standards, as state-granted variances seem to be easily extended.

Overall, the institutional setting provides little incentive for increasing the economic efficiency of water use, for economically managing the TDS level of the waters of the Upper Colorado Basin, or for achieving an optimal level of air quality.

Integrated Economic, Hydro-Salinity, and Air Quality Models of the Upper Colorado River Basin

There are three main elements in a set of these models:

1. a regional input-output (I-O) model of the regional economy;
2. hydrologic models which trace monthly surface and groundwater flows and water uses for drainage basins within the economic region; and
3. air dispersion models to trace gaseous residuals discharges from point and diffuse urban sources and to estimate the resulting ambient air quality.

The linkage among the three types of models is the set of regional industrial activity levels (total gross output of each industry, TGO) which are projected into the future under alternative assumptions about technology, price, and policy variables. The I-O table assures that the projected industry levels are consistent with one another and that the indirect effects of any particular project or development on other industries are picked up. The TGO levels from the I-O projections are inputs into the hydro-salinity and air quality models where, in conjunction with water use and residuals generation coefficients, they determine the demands for water, TDS added to the return flows, and the amounts of particulate gaseous residuals generated.

I-O models have been constructed for the rather large subbasins of the Green River, the Upper Main Stem of the Colorado River, and the San Juan River which together constitute the Upper Basin of the Colorado River. (See figure 4-1.) The hydro-salinity models can be calibrated to any set of mutually exclusive and exhaustive drainage basins within a subbasin, the only constraints being the availability of hydrologic and water quality data and the ability to disaggregate the projected TGOs by drainage basin in a sensible way. For the initial modeling project, the subbasins were broken down as follows for hydrologic modeling: (1) Upper Main Stem--three drainage basins; (2) Green River Basin--six drainage basins; and (3) San Juan Basin--three drainage basins.

The air quality model takes the TGOs and maps them into a specified set of point and diffuse sources of particulates, sulphur dioxide, nitrogen oxide, carbon monoxide, and unburned hydrocarbons. The outputs of the model are both tabular forms of grids of concentrations and isopleth maps. These are illustrated later in the paper.

The I-O models were originally constructed from survey data gathered in 1963 as part of a large-scale study of the emerging mineral quality problem in the Colorado River. The coverage of the survey was essentially 100 percent for major, large-firm industrial activities, down to 15 percent or so for small-scale commercial activities.[5] The effort required about eighteen months and cost $500,000 (including the three I-O models for the Lower Basin). Thus the models are over ten years old, certainly a potential source of trouble for analysis for environmental quality management in a rapidly growing region. The sectors contained in the three subbasin I-O models are shown in table 4-2. While the I-O models cover subbasins of the Upper Colorado River Basin, their imports and exports are simply divided between those from or to other subbasins and from or to the rest of the world. Particular interbasin flows are not indicated, so the set of models does not constitute a true interregional I-O model of the entire basin.

Table 4-2. The Sector Composition of the I-O Models for the Three Subbasins of the Upper Colorado River Basin

Upper Main Stem

1. Range Livestock
2. Feeder Livestock
3. Dairy
4. Food & Field Crops
5. Truck Crops
6. Fruit
7. Forestry
8. All Other Agriculture
9. Coal
10. Oil & Gas
11. Uranium
12. Zinc
13. All Other Mining
14. Food and Kindred Products
15. Lumber & Wood Products
16. Printing & Publishing
17. Fabricated Metals
18. Stone, Clay & Glass
19. All Other Manufacturing
20. Wholesale Trade
21. Service Stations
22. All Other Retail
23. Eating & Drinking Places
24. Agricultural Services
25. Lodging
26. All Other Services
27. Transportation
28. Electric Energy
29. Other Utilities
30. Contract Construction
31. Rentals & Finance
32. State & Federal Govts.
33. Local Government
34. Wages (Households)
35. Profit & Other Income (Households)
36. Depreciation Allowance
37. Inside CRB Imports
38. Outside CRB Imports
39. Inventory Depletions
40. Total Gross Outlays

Green River

1. Agriculture
2. Forestry
3. Oil & Gas
4. Coal
5. Uranium
6. Food & Kindred Products
7. Lumber & Wood
8. Printing & Publishing
9. Stone, Clay & Glass
10. All Other Manufacturing
11. Wholesale Trade
12. Service Stations
13. Eating & Drinking Places
14. All Other Retail
15. Oil Field Services
16. Lodging
17. Other Services (except professional)
18. Transportation
19. Electric Energy
20. All Other Utilities
21. Contract Construction
22. Rentals & Finance
23. State & Federal Govts.
24. Local Government
25. Wages (Households)
26. Profits & Other Income (Households)
27. Inventory Depletions
28. Depreciation Allowances
29. Inside CRB Imports
30. Outside CRB Imports
31. Total Gross Outlays

Table 4-2 (continued).

San Juan River

1. Range Livestock
2. Dairy
3. Field Crops
4. Fruit
5. Forestry
6. All Other Agriculture
7. Coal
8. Oil and Gas
9. Uranium
10. All Other Mining
11. Food & Kindred Products
12. Lumber & Wood Products
13. Printing & Publishing
14. Stone, Clay & Glass
15. All Other Manufacturing
16. Wholesale Trade
17. Service Stations
18. All Other Retail
19. Eating & Drinking Places
20. Agricultural Services
21. Oil Field Services
22. Lodging
23. All Other Services (except professional)
24. Transportation
25. Electric Energy
26. Other Utilities
27. Contract Construction
28. Rentals and Finance
29. State & Federal Govt.
30. Local Government
31. Wages } Households
32. Profits & Other Income } Households
33. Inventory Depletions
34. Depreciation Allowances
35. Inside CRB Imports
36. Outside CRB Imports
37. Total Gross Outlays

The logic of the hydro-salinity model is indicated in figure 4-2. Roughly, the hydro-salinity model contains the following features on a monthly basis:

1. main stem inflows of water and TDS;
2. tributary inflow based on two sample "key streams" which represent precipitation or snowmelt conditions in the drainage basin except for the valley bottom;
3. reservoir storage and operating rules for releases from storage "high" in the basin;
4. municipal and industrial surface and groundwater withdrawals and return flows, including pick-up of TDS;
5. agricultural surface diversions, canal losses, field applications, root zone storage, evapotranspiration; deep percolation; TDS pick-up, and lagged return flows;
6. phreatophyte and general evaporative losses;
7. imports to and exports of water from the drainage basin;
8. groundwater stocks for the basin;
9. end of basin reservoir storage and release rules; and
10. surface and groundwater flows out of the basin.

The model is driven by monthly time series of main stem and tributary inflows, precipitation, and temperature, in the form of either historical data series or synthetically generated series. Hydrologic basins can be run in sequence, so that each basin's output becomes the input for the appropriate lower basin. In this way, the behavior of the entire Upper Basin can be simulated simultaneously.

The data needs of this type of model can be broken down into: (1) data for purposes of calibration or estimating the parameters of the model;

Figure 4-2. Flow Diagram of the Upper Main Stem Hydro-Salinity Model

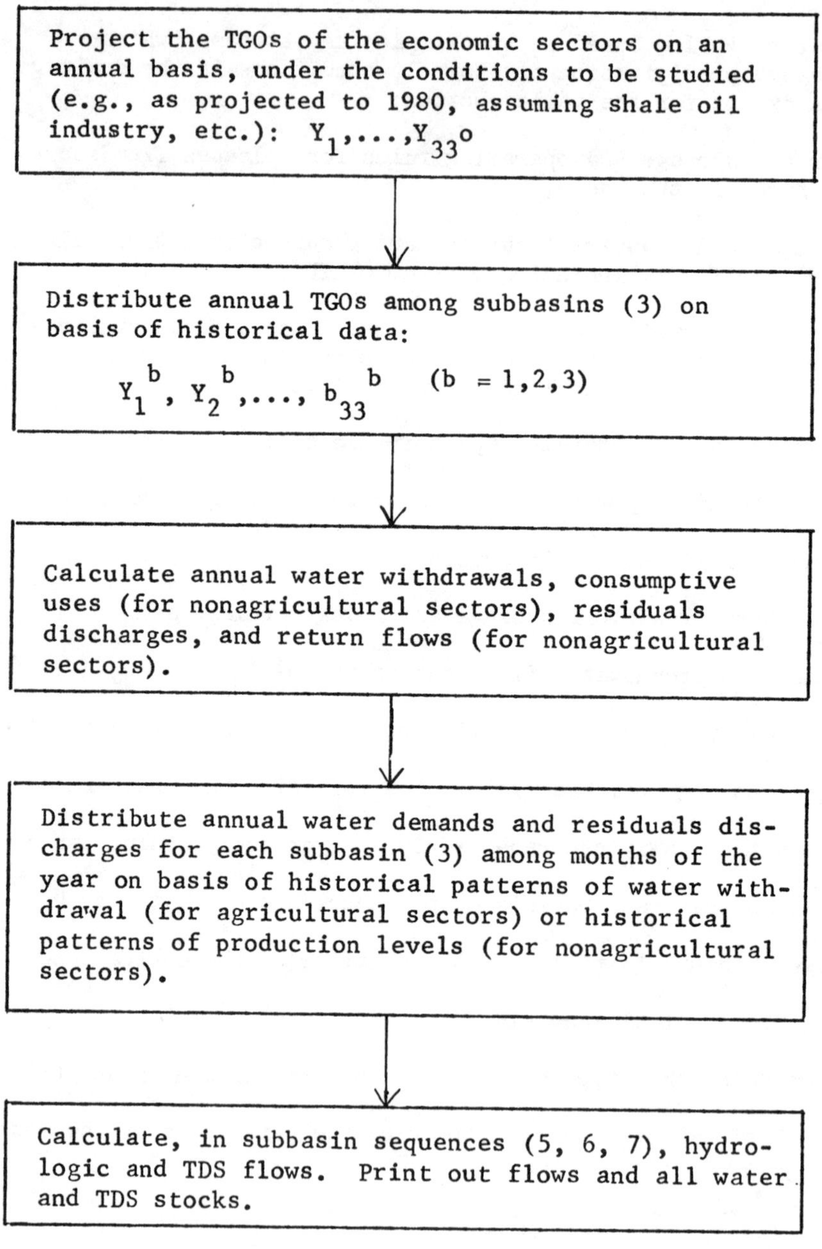

Project the TGOs of the economic sectors on an annual basis, under the conditions to be studied (e.g., as projected to 1980, assuming shale oil industry, etc.): $Y_1, \ldots, Y_{33}^o$

↓

Distribute annual TGOs among subbasins (3) on basis of historical data:

$$Y_1^b, Y_2^b, \ldots, b_{33}^b \quad (b = 1,2,3)$$

↓

Calculate annual water withdrawals, consumptive uses (for nonagricultural sectors), residuals discharges, and return flows (for nonagricultural sectors).

↓

Distribute annual water demands and residuals discharges for each subbasin (3) among months of the year on basis of historical patterns of water withdrawal (for agricultural sectors) or historical patterns of production levels (for nonagricultural sectors).

↓

Calculate, in subbasin sequences (5, 6, 7), hydrologic and TDS flows. Print out flows and all water and TDS stocks.

and (2) data for purposes of "driving" the model to ascertain the hydro-salinity consequences of a certain set of economic conditions, as specified by the TGOs.

For calibrating the model, it was necessary to have the following data:

1. the flows of various tributaries to find appropriate key streams;
2. the gaged main stem river flow at each point considered to be the end point of a subbasin;
3. climatological data on precipitation and temperature, measured at "representative" weather stations on the valley floor of each subbasin;
4. data on reservoirs within the region, including dates of completion and filling, and operating rules;
5. data on withdrawals, consumptive uses, and residuals discharges in nonagricultural sectors; and
6. data on cropping patterns so that appropriate values for evapotranspiration can be calculated.

The calibration phase is naturally data intensive, and it is often difficult to find historical records for appropriate stream gaging and weather stations which overlap for a sufficiently long period to permit reliable estimates of the parameters to be made. In the present case, the period 1944-1968, a twenty-five-year period, was used for calibration. Fortunately, it was possible to get a twenty-five-year overlap for the hydrological and meteorological variables (key streams, main stem outflows, precipitation, and temperature). TDS concentrations were less well recorded, and different periods of record had to be used to calibrate the TDS equations. The hydrologic models for each subbasin were first calibrated, then the salinity calibrations were undertaken.

Following calibration of the models, application to the analysis of the implications of a particular set of economic conditions requires again that the models be "driven" by some set of time series of the hydrology and salinity concentrations of the key streams, precipitation, and temperature. Thus far, only the data of record are available to drive the model, but experiments have been carried out to test the synthetic record approach. The major advantage of that approach is that many years of consistent hydro-meteorological data can be run through the model for given economic conditions, yielding not merely averages of system performance but <u>well-described frequency distributions of the statistics of the system</u> (e.g., distributions of monthly flows, salinity concentrations at different points, frequencies of water shortage). Typical tabular output from the application of the hydro-salinity model is shown in tables 4-3 and 4-4. Plotted output is also available.

The APGDM[6] is a computer program designed to be used in conjunction with the I-O economic models of the Colorado River Basin. The program computes airborne residuals generation and discharge rates and ground-level air quality patterns on a quarterly basis, corresponding to the level of economic activity determined either by field surveys or by projections of the I-O Model. To make the modeling effort tractable in the face of present data inadequacies, certain simplifying assumptions are made which limit the accuracy of the model. However, although calibration data for the APGDM are sparse, the available data indicate that the model predicts ground level air quality concentrations to an accuracy of $\pm$ 20 percent.

Table 4-3. Water Statistics Table for Basin No. 5
Eighteen-Year Run Under 1960 Conditions

	Mean Flow 10^3 af	Standard Deviation 10^3 af	Variance 10^6 af	Alpha 3 (skew)	Maximum Flow 10^3 af	Minumum Flow 10^3 af	High Diversions	High Exports
January	82	6.3	40	0.11	94	70	0	0
February	72	6.9	48	-0.89	85	56	0	0
March	86	11.8	138	0.08	110	64	0	0
April	134	35.2	1,240	0.88	211	87	0	0
May	430	157.3	24,760	1.04	793	268	0	0
June	759	241.0	58,070	-0.47	1,152	228	0	0
July	358	227.2	51,610	1.47	998	124	0	0
August	162	69.0	4,760	1.08	325	77	0	0
September	116	38.4	1,470	1.22	220	74	0	0
October	100	28.2	798	1.32	183	64	0	0
November	95	10.0	100	0.63	117	81	0	0
December	89	6.8	46	0.82	102	81	0	0

Table 4-4. TDS Statistics Table for Basin No. 5
Eighteen-Year Run Under 1960 Conditions

	Mean 10^3 tons	Standard Deviation 10^3 tons	Variance 10^6 tons	Alpha 3 (skew)	Maximum Load 10^3 tons	Minimum Load 10^3 tons
January	51	5.5	29.8	2.75	70.9	45.7
February	47	4.0	15.7	2.28	60.9	42.4
March	51	5.2	27.3	1.16	65.9	44.9
April	65	11.2	125.4	0.58	85.9	47.8
May	124	26.9	725.4	0.85	184.1	93.5
June	173	35.5	1262.5	-0.94	222.7	84.6
July	113	46.9	2197.1	1.15	241.7	56.9
August	84	29.7	884.3	0.61	148.6	41.2
September	65	12.5	155.7	-0.04	87.9	45.2
October	54	8.5	72.7	0.82	75.8	39.7
November	54	3.6	12.9	0.48	59.6	48.6
December	52	3.1	9.9	0.10	57.4	47.5

The model was also designed to predict, for comparative purposes, future air quality levels resulting from the simultaneous effects of economic growth in the region, improved gaseous residuals discharge reduction techniques, and more strict state and federal gaseous residuals discharge standards.

Section 110(a)(1) of the Federal Clean Air Act of 1970 required that each state prepare implementation plans outlining in detail how ambient air quality standards promulgated by the Administrator of the Environmental Protection Agency would be met by certain time deadlines. As a result, the states were required to determine the pattern of gaseous residuals discharges within their boundaries. It is this survey, the discharge inventory, which provides the APGDM with its largest single source of input data. As the quality of the discharge inventory varies from state to state, so does the usefulness and reliability of the APGDM.

Assuming the existence of a discharge inventory, residuals discharge factors may be applied to the industrial, municipal, and domestic activity levels to determine how much residuals discharge may be attributed to each source on a seasonal basis. Once the residuals discharge rates are known, the model predicts the dispersion patterns of these residuals in the atmospheric surface layer based upon historical meteorological data and known dispersion characteristics of the atmosphere.

The I-O model's geographical subdivisions are also satisfactory for analyzing dispersion of gaseous residuals because air drainage patterns and water drainage patterns in mountain valleys are analogous. In addition, gaseous residuals do not disperse from one subbasin to another because the subbasins are separated by high mountain ridges in the Colorado River Basin.

Two types of sources of gaseous residuals exist. A <u>point source</u> is a single, stationary source of significant discharge of one or more gaseous residuals. Point sources are usually industrial in nature, having one or more large stacks. An <u>area source</u> includes all sources of gaseous residuals, mobile or stationary, in some defined area, which sources are too small or numerous to be individually inventoried. Area sources include such activities as residential space heating, commercial space heating, and automobile travel.

The dispersion of a gaseous residual from a stack depends not only on the characteristics of the discharge, i.e., exit velocity, volume, temperature, but also on the meteorological conditions at the site. Wind velocity and direction, ambient temperature, and mixing depth are all important data which must be known if a reliable air quality grid is to be constructed. Wind velocity and direction are usually presented in a wind rose from which the percentage of time during which the wind blows at a given speed from a given direction may be determined. Historical wind roses and ambient temperatures are available for the larger urban areas and some airport sites in the Upper Main Stem.

Ground level air quality is presented in two ways. The first is in the form of a grid in which the ground level concentration of a residual is given in micrograms/cubic meter at each grid point. The grid points are located on lines along the eight primary compass directions at equal incremental distances from the source. The second display method shows lines of constant concentration (isopleths) in an area. Two isopleth maps are shown as figures 4.3 and 4.4.

Figure 4-3. Upper Main Stem Subbasin, Grand Junction Zone
Suspended Particulate Isopleth Map, First Quarter 1970

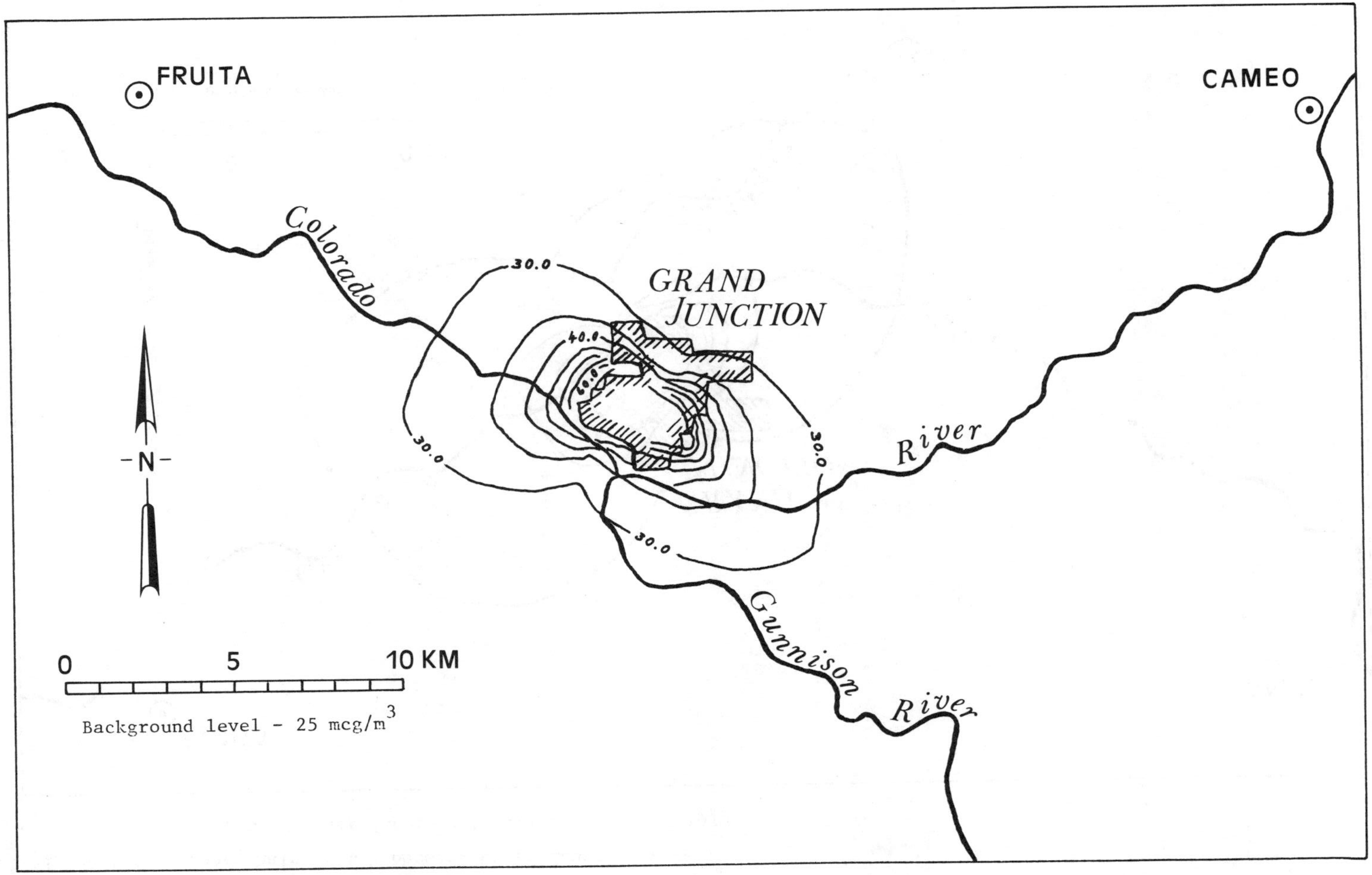

FRUITA
CAMEO
GRAND JUNCTION
Colorado River
Gunnison River
30.0
40.0
N
0 5 10 KM
Background level - 25 mcg/m^3

Figure 4-4. Upper Main Stem Subbasin, Grand Junction Zone
Sulphur Dioxide Isopleth Map, First Quarter 1970

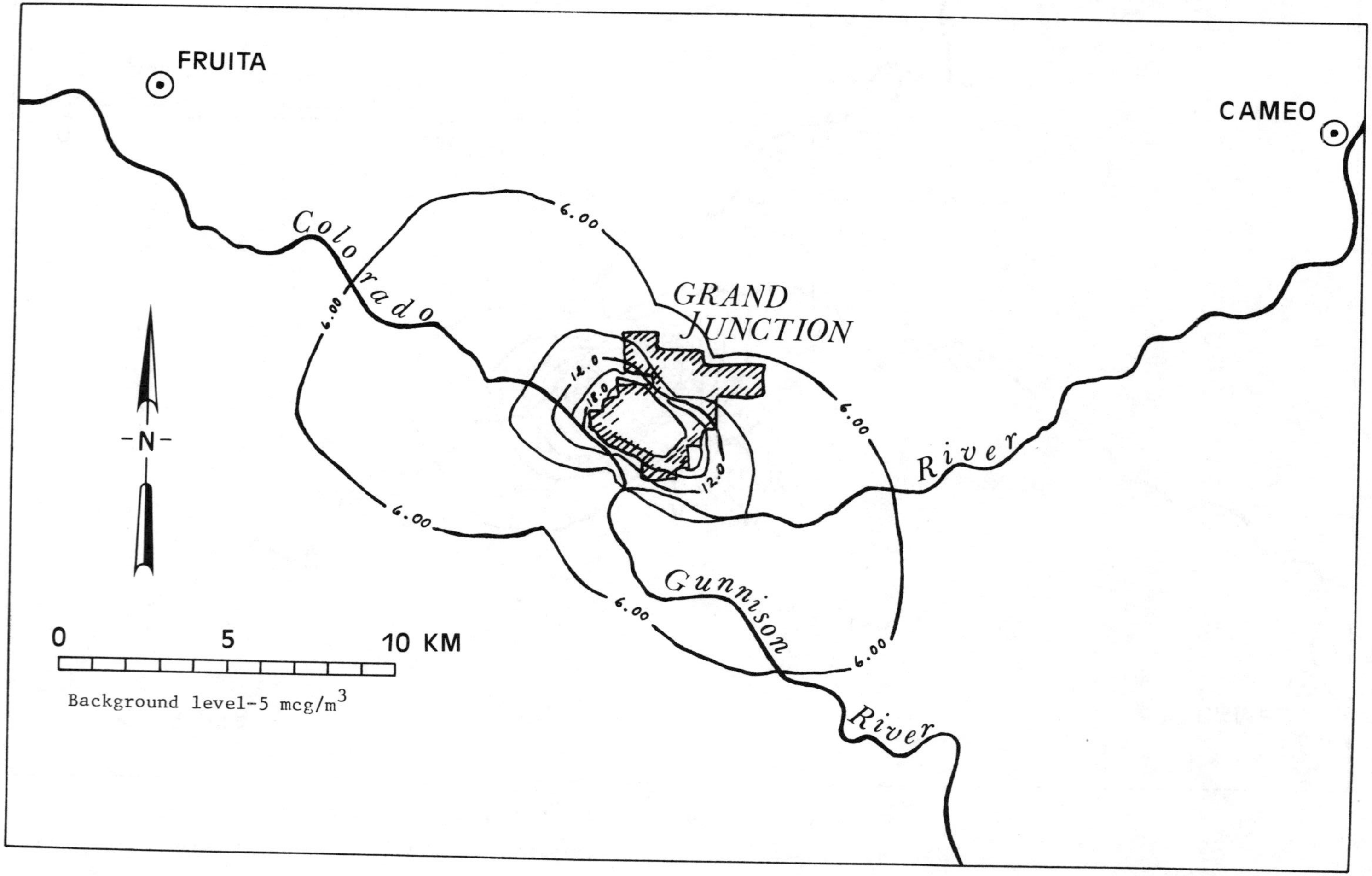

The program for computing ambient air quality has two major sequential steps. The first computes residuals discharge rates by applying discharge factors to source data contained in the source file. Each industrial process has unique discharge factors for the following five residuals: airborne particulates ($d<20\mu$); SO_2 - sulphur dioxide; NO_x - nitrogen oxides; CO - carbon monoxide; unburned hydrocarbons. In addition to discharges from industrial processes, discharges from fuel combustion--from facilities auxiliary to the industrial processes and from home heating--are computed. The arithemetic sum of process and fuel combustion discharges from point and area sources are inputs to the second half of the program.

The dispersion[7] section of the program uses the well-known atmospheric diffusion model of Gifford to transform regional source discharge patterns and meteorological data for a given quarter into estimated ground level, quarterly averaged concentrations. Certain options are available to the user in that several atmospheric stability classes may be selected. (The less stable the atmosphere, the more rapidly gaseous residuals disperse.) In any given day two, three, or four of the six recognized stability classes may exist; hence it is not proper to speak of an "average" stability class on a quarterly basis. The dispersion sector of the program computes air quality levels which may be expected on the average if a given stability class were to persist for a long time period.

Current Applications

The first application of the set of models was to investigate the economic, hydrologic, and water quality effects of a substantial reduction in irrigated agriculture in the Upper Main Stem subbasin. A rational program of acreage reduction would phase out economically marginal lands first, especially those that also contribute heavily to the TDS load. It was simply assumed, however, that average units in the food and field crops sector and in the range cattle sector would be phased out: $10 million per year of the TGO of the former, and $40 million per year of the TGO of the latter, an overall acreage reduction of 416,000 acres. It was further assumed that these direct decreases in output would take the form of reductions in exports from the Upper Main Stem subbasin rather than reductions in deliveries to other industries in the subbasin.

The 1980 base projections and the direct and indirect reductions in TGOs are shown in table 4-5, as derived from the I-O Model. The sum of reductions in all TGOs was $107 million per year, and the total reduction in income payments to households in the region--including local profits, rents, interest--was $56 million per year.

To determine the hydrologic and salinity effects of the reduction in agricultural activities, the hydro-salinity model was run twice: once using the 1980 base TGOs (column 1, table 4-5) and again using the reduced TGOs (column 3, table 4-5). The same 25 years of historical hydrologic data were used to drive the model in both cases. The two corresponding sets of flow and TDS data are shown in tables 4-6 and 4-7. The annual water saved directly and indirectly amounted to about 0.6 maf and the annual reduction in TDS was about 1.2×10^6 tons. Thus for the losses in basin income, we "buy" the joint product of additional water released for other in-basin and out-of-basin uses and a reduction in TDS loading. If we "cost-out" TDS reductions by alternative methods--such as by reduction in TDS discharges from

Table 4-5. Upper Main Stem Economy for 1980 with Reduced Agricultural Activities

Sector	Total 1980 Gross Output, millions of dollars		
	Initial Projection	Total Reduction	New Projection
1	61.8	45.2	16.6
2	8.5	0.7	7.8
3	4.4	0.3	4.1
4	10.9	10.2	0.7
5	1.2	<0.1	1.2
6	4.3	0.3	4.0
7	9.8	<0.1	9.8
8	1.4	0.1	1.3
9	15.9	0.3	15.6
10	4.8	0.0	4.8
11	82.9	0.0	82.9
12	20.7	0.0	20.7
13	22.0	0.1	21.9
14	29.7	2.7	27.0
15	10.6	0.1	10.5
16	6.4	5.9	0.5
17	4.5	<0.1	4.5
18	5.2	0.1	5.1
19	58.2	2.4	55.8
20	42.7	2.4	40.3
21	5.7	1.1	4.6
22	82.9	8.9	74.0
23	35.0	1.6	33.4
24	5.3	2.0	3.3
25	51.9	0.1	51.8
26	80.0	2.9	77.1
27	55.6	3.5	52.1
28	23.9	1.6	22.3
29	27.1	2.1	25.0
30	128.0	3.2	124.8
31	108.0	9.4	98.6
	1009.3	107.2	902.1

Table 4-6. Upper Main Stem Outflows of Water and TDS Under Projected 1980 Conditions

Month	Mean Flow, 10^3 Acre-Feet	TDS Load, 10^3 Tons
January	194	284
February	182	277
March	231	308
April	318	288
May	660	369
June	956	1,025
July	480	898
August	315	683
September	340	565
October	241	320
November	259	326
December	257	343
Average Month	(369)	(474)
Annual Total	4,433	5,686

Table 4-7. Upper Main Stem Outflows of Water and TDS After Phasing Out Selected Agricultural Activities, 1980 Conditions

Month	Mean Flow 10^3 Acre-Feet	TDS Load, 10^3 Tons
January	197	284
February	185	278
March	234	309
April	366	305
May	977	425
June	1,179	625
July	549	557
August	301	409
September	307	381
October	236	280
November	246	305
December	252	331
Average Month	(419)	(374)
Annual Total	5,029	4,489
Differences in flow and TDS with and without selected agricultural activities	+596	-1,197

important point sources--at \$20 per ton of reduction, and impute the remaining \$32 million as the cost of getting 0.6 maf of water annually, the cost is approximately \$54 per acre-foot of water. If we cost out the salt reductions at a cost more nearly representative of the prospective costs of the proposed desalting plant to reduce salinity in the Wellton-Mohawk return flow (using an annual cost of $\$10 \times 10^6$ to remove 250,000 tons per year), a figure of approximately \$40 per ton removed, the implied cost of the water saved drops to \$14 per acre-foot.

However, the income losses have been biased upward for reasons detailed earlier. Further, all possible increases in efficiency which would conserve water and reduce TDS discharges without proportional income losses have not been considered in the analysis. Finally, since any form of compensation for farm land or water rights would in fact result in large capital payments to the sellers of water rights, it could be expected that some portion of this capital would be reinvested in the Basin, raising incomes once again towards their original levels. The actual costs would, in the final analysis, be much lower than the \$56 million indicated above.

There is, thus, striking empirical evidence that reductions in irrigated agricultural acreage would have highly beneficial effects in terms of water saved and reductions in TDS load, at levels of regional income costs which are likely to compare very favorably with alternative solutions to the water quantity and quality problems. In this case, the implied changes in air quality were negligible.

A second application of the set of models has been in the investigation of the effects of various levels of expansion of coal mining in the valley of the North Fork of the Gunnison River. This expansion, which is already underway, may take either of two forms: (1) a large increase in coal production for export from the area; (2) a large increase in coal production, accompanied by large power plants located in the same area to provide electric energy for shale oil developments to the north. In the case of power plants locating in the area, power plant location will be critical because of the mountain valleys involved. Thus, the economic, water quantity, water quality, and air quality effects of alternative levels of coal output, with and without energy generation, with alternative power plant sites are currently being investigated. The results of that analysis are not included in this paper.

A third application is to investigate the direct and indirect economic effects of other alternative means of reducing TDS in the Colorado River. The steps being investigated include: (1) costs of failure to prevent salinity increases; (2) regional impacts of increased irrigation efficiencies and other changes in agricultural practices. This project involves several universities and the use of additional analytical tools.

A fourth application is to study the limits to the various new energy industries in the Upper Basin (coal gasification, shale oil, and more thermal-electric energy generation) and the possible trade-offs among them in the light of water availability and various constraints on water quality. Possible trade-offs against other types of economic activity will also be studied.

Difficulties Encountered

One difficulty in regional analysis of this type is keeping the economic I-O model effectively updated. The basin being studied is still relatively sparsely developed and was more so the ten-plus years ago when the model was assembled. Relatively small amounts of growth or the movement of a few firms can change the coefficients quickly and significantly. Surveys sufficient to guarantee accurate updated I-O models are costly and time consuming.

Other difficulties involved getting data on water quality and on gaseous residuals discharges. The streams of the southwestern United States are only sketchily gaged for water quality. As the area being modeled gets smaller, it becomes increasingly difficult to find the required water quantity and quality data. The inventories of gaseous residuals discharges were carried out by the states; some did their work well and promptly, others poorly and behind schedule.

The greatest difficulty is that as a University research group and even though our work has been sponsored by the Economic Development Administration of the federal government, we have not been able to establish a working linkage with the state governments and federal agencies which are really making the development decisions.

References and Footnotes

[1] Charles W. Howe and K. William Easter, Interbasin Transfers of Water: Economic Issues and Impacts (Baltimore, Johns Hopkins University Press for Resources for the Future, 1971) table 30.

[2] U.S. Environmental Protection Agency, Regions VIII and IX, The Mineral Quality Problem in the Colorado River Basin (1971).

[3] Current--at this point and elsewhere in the paper--refers to conditions around 1970.

[4] The Four Corners Area is so called because it includes parts of the four states--New Mexico, Arizona, Utah, and Colorado--which have one point in common.

[5] For more detail on model construction, see Bernard Udis (ed.), "An Interindustry Analysis of the Colorado River Basin With Projections to 1980 and 2010", Bureau of Economic Research, University of Colorado, 1968 (prepared under Contract No. WA67-4 between the U.S. Department of Interior and the University of Colorado).

[6] APGDM = air pollution generation and discharge model.

[7] Frank A. Gifford, Jr., The Problem of Forecasting Dispersion in the Lower Atmosphere, Weather Bureau Research Station, Oak Ridge, Tennessee (1961).

Chapter 5

AN ANALYSIS FOR REQM IN THE LJUBLJANA, YUGOSLAVIA, AREA

Daniel J. Basta, Blair T. Bower, and James L. Lounsbury

The study reported herein describes an analysis for residuals-environmental quality management (REQM) in the metropolitan area of Ljubljana, Yugoslavia. Previously the REQM approach had been applied mainly to highly developed areas, such as the Lower Delaware Valley of the United States,[1] where substantial amounts of data existed, and substantial financial and analytical resources enabled use of sophisticated computer modeling techniques. Application to such areas still left the questions: to what extent can a comprehensive REQM analysis be made operational and what relevant questions for REQM can be answered in areas where: (1) the economic, political, and social context is different than in the United States; (2) resources and data for analysis are limited; and (3) economic and political structures are not clearly defined or are non-existent with respect to responsibility for environmental quality management? To attempt to answer that question, the REQM analysis of the Ljubljana, Yugoslavia area was initiated in 1973.[2]

That the study was undertaken in Ljubljana, Yugoslavia was a result of the conflation of two factors. The first was the existence during the period 1966-1976 of a joint U.S.-Yugoslav program of research on urban problems in the Urbanisticni Institut of the Socialist Republic of Slovenia. This provided the institutional base for the study. The second was the existence of a program of research on regional REQM at Resources for the Future, Washington, D.C. since about the same time. This provided the analytical base for the study. The study was funded primarily by grants of U.S. P.L.-480 funds to the Johns Hopkins University Center for Metropolitan Planning and Research, supplemented by a small grant from Resources for the Future. Support for preparation of the final report was also provided by Resources for the Future.

The project was organized to be as Yugoslav-oriented as possible, in order that the results would be useful. This meant attempting to maximize: the participation of Yugoslav staff members; the development and utilization of data specifically reflecting Yugoslav conditions; and the development and analysis of REQM strategies which were perceived by the Yugoslavs as being relevant to their conditions.

REQM in the Yugoslav Context

The political-economic-social structure of Yugoslavia differs substantially from that of the United States. Despite the many differences in each society's demand for goods and services, resources, the distribution of economic services, residuals generation and discharge, the perceptions of and attitudes toward environmental quality and governmental structure, many commonalities in the two contexts are evident. Three basic conditions exist in both contexts if ambient environmental quality (AEQ) is to be improved. One, some activities must do some things differently than they did before. Two, there is always some initial cost, no matter whether or not the cost diminishes or becomes negative in the future--that is, returns from the sale of recovered materials and/or energy more than offset costs to reduce discharges of residuals. Three, regardless of who is requested to pay the costs, the initial response is almost always negative. Although the characterization of the typical situation is not profound, the conditions enumerated suggest that the critical link in the process of identifying, analyzing, and implementing physical measures to improve AEQ lies in the successful development of "believable" minimum cost REQM strategies that are compatible with the existing overall institutional philosophy of a society. Compatible means that they can be implemented through some relatively simple incentive measures involving low administrative costs, an uncommon combination in any society.

Despite the similarities between the two countries with respect to REQM, there are important differences. Whereas the United States is industrialized--and presumably moving into the "post-industrial" society--Yugoslavia is a rapidly industrializing country where industrial development and the demand for goods and services did not reach significant levels until the 1960-1975 period. This means that the resultant increase in residuals generation and discharge has only recently produced noticeable and perceived adverse impacts on AEQ. However, the emphasis on economic growth and increased standard of living has, for the most part, not included a similarly expanded program for the management of the resulting residuals. It is not surprising that until very recently legal provisions concerning environmental quality have been seriously lacking at all levels of government in Yugoslavia. But over the past decade, as citizens and planners have become increasingly aware of deteriorating AEQ and the effects thereof, some significant institutional changes have occurred to alter this situation. Nevertheless, it is likely that it will be several years before Yugoslavia's organizational structure for, and the directions of, REQM crystallize.

Goals of the Study

Given the above background, the following were the goals of the study. One goal was to attempt to apply the REQM approach to, and to test its utility in, a specific "real-world" context in which it was likely that many of the data for analysis of management strategies did not exist. A second goal was to assess the utility of applying, and the differences of so doing, a methodological approach developed in the United States to another socio-economic-political system. The hypothesis was that the methodology itself is _apolitical_--the institutional context and constraints are not. A third goal was to generate useful and meaningful information on the implications of alternative REQM strategies for the specific area of Ljubljana, useful in terms of actual decisions with respect to allocation of resources to REQM in the Yugoslav context, meaningful in relation to Yugoslav organizations and decision makers. A fourth and final goal was to stimulate Yugoslav agencies and organizations to take an active role in the study so that they could: (a) insure that the analysis was as Yugoslav-oriented as possible; and (b) eventually expand upon and carry out the work initiated by the project.

Location and Characteristics of the Study Area

The study area is comprised of five communes--Centar, Siska, Vic, Moste, and Bezigrad, covering 903 square kilometers (350 square miles),[3] and is situated in a basin along the southern slopes of the Julian Alps in Slovenia, the northernmost republic of Yugoslavia (see figure 5-1). The five communes comprise about 90 percent of the basin's land area. In the center of the basin is the city of Ljubljana and its suburbs. The urbanized area--comprised of all Centar commune and parts of the other four communes--includes about

Figure 5-1. Location of the Study Area

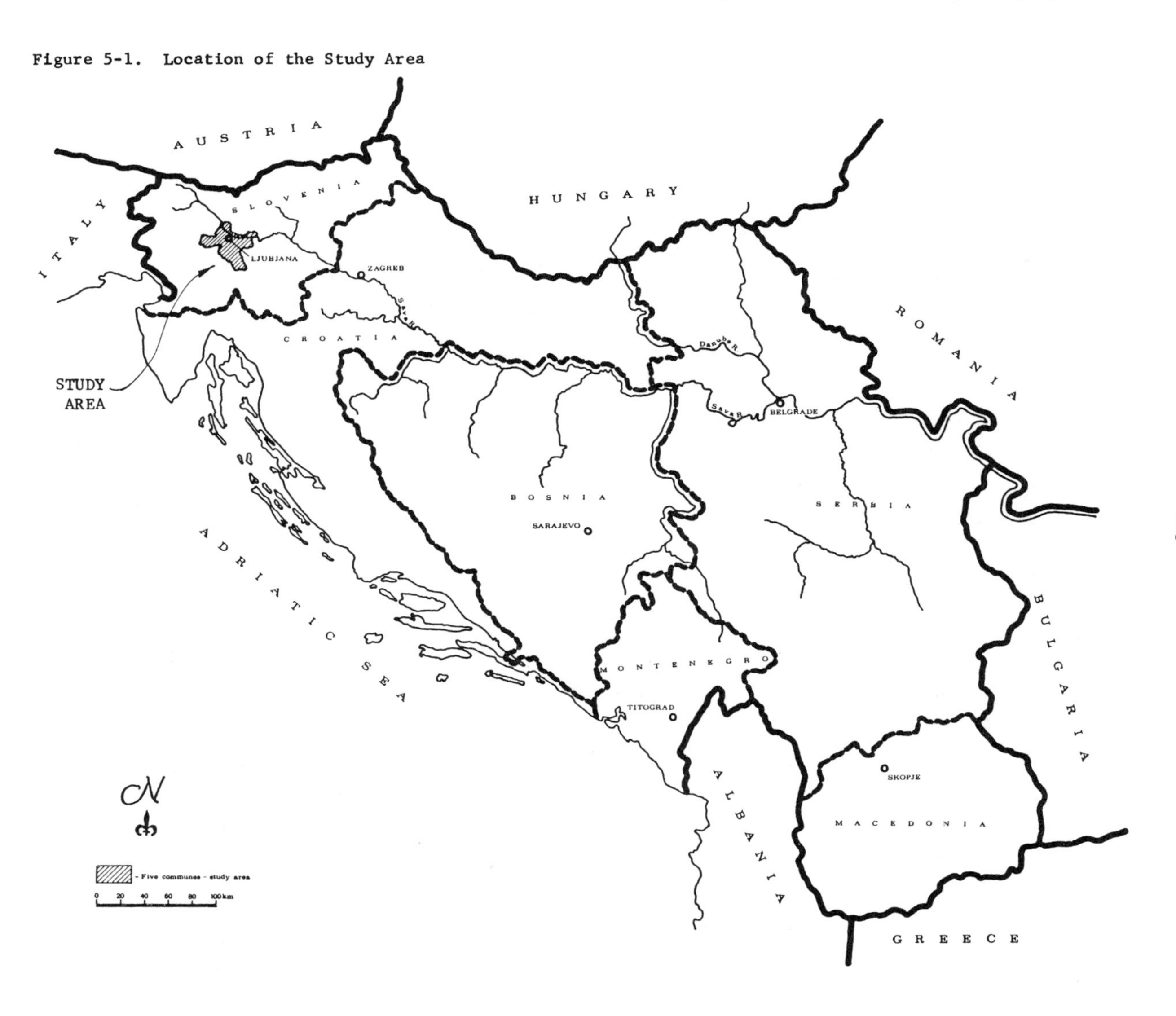

5 percent of the land area of the basin, and in 1972 contained about 70 percent of the population of the five communes. To the south of the city there is a large area of poorly drained, high water table land--the moor area; to the north lies a dry alluvial plain in which the remaining 30 percent of the population of the five communes resides. The mountains bordering the basin are a major center for alpine sports. Much of the population engages in such activities as skiing and hiking and consequently is very much aware of the value of the natural resources making these activities possible.

Air Resources

The meteorological characteristics of the basin result in a nearly closed system with respect to air quality. The general climate of the area is classified as temperate. Periods of fog, temperature inversion and limited ventilation occur which are most severe in winter and often last seven days, and occasionally longer. The ambient quality of the relatively stationary air masses during such periods is often poor, because of the accumulation of gaseous residuals discharged from activities within the basin.

Water Resources

Both ground and surface water are used for water supply in the area. The water for the Ljubljana city system is withdrawn from ground water aquifers in the recharge area north of the city. Most of the water which is used in industrial operations is withdrawn from local surface sources. Users outside the water supply area not supplied by the city water system depend on individual wells tapping the ground water aquifer.

Two major rivers flow through the basin. The Sava, the largest river in Yugoslavia, flows southeast in Slovenia through most of the country to its confluence with the Danube at Belgrade. Just east of Ljubljana the Sava is joined by the Ljubljanica, which flows eastward through the center of the city. Most of the liquid residuals from the activities in the study area are discharged directly (unmodified) into these two rivers, or one of their several smaller tributaries. Water quality of the Ljubljanica is very poor, but because of the rapid rate of flow and substantial reaeration, water quality of the Sava in the vicinity of the study area is generally acceptable.

Population and Employment

In 1972, the population of the five-commune area was about 257,000,[4] an increase of 25 percent over the 1962 population. During this period a significant shift in the distribution of economic activity occurred. Major new production activities initiated in the area included industry--primarily metal processing, electrical products, and food processing--and commerce, including a significant amount of tourism in the summer. Agriculture declined in importance relative to other activities. In addition, much of the agricultural activity is private and unmechanized, with little use of chemical fertilizers and pesticides. In 1972 the total work force comprised about 105,000 individuals.

Institutional Arrangements

In Yugoslavia, the communes have the responsibility for environmental quality management. At that level of government no single agency exists with responsibility for environmental quality decision making or for REQM. However, there are four institutions which can influence REQM strategies in the five-commune area.

One is the environmental council of the city (urbanized area). This is a non-technical advisory group which provides advice on environmental issues to the city assembly. A second is Splosna Vodna Skupmost, the technical arm of the Slovenian Republic Water Quality Commission for the Upper Sava River Basin. Its functions comprise research and provision of information to the republic assembly, concerning intercommunal water quality problems and policies in the Upper Sava River Basin. A third is SEPO, a non-governmental credit-providing agency affiliated with Ljubljanska Banka,[5] which has adopted the policy of not approving credit for projects with potentially negative AEQ impacts. The fourth is the Republic Committee for Air Quality, which is an advisory group of professional and technical people. The function of the Committee is to formulate legislation at the republic level relating to air quality management. All four institutions are based in Ljubljana, the center of the study area. The first and third institutions have responsibilities across the total range of environmental issues; the second only with respect to water; the fourth only with respect to air.

Conditions for Analysis and Study Outputs

Conditions for Analysis

An important step in setting up an REQM study is to define the total set of conditions under which the analysis will be performed. Although it is not usually possible to define accurately, at the outset of a study, all the conditions under which a study will be carried out, a necessary initial step is to define them as completely as possible. Analysis conditions comprise the decisions that must be made concerning, for example: the boundaries of the study area; the methodology used for analyzing REQM systems; what variables are assumed fixed; what time period is to be covered; and how to allocate the available analytical resources, that is, the time, money, personnel and computational facilities available for the study.

Five general conditions affected the organization and content of the study. They were: (1) limited resources--funding, personnel, reference materials; (2) lack of a "clear-cut" definition of the REQM problem by a user agency, because no single, unified agency responsible for REQM in the area existed; (3) eighteen-month time period for performing the analysis; (4) political constraints on data accessibility and data collection; and (5) an almost complete lack of readily usable data on residuals generation and discharge. In addition to these general conditions, the following eight more detailed analysis conditions were specified for the study:

1. Residuals generation coefficients used be developed to reflect Yugoslav conditions;
2. REQM strategies developed be the result of a joint Yugoslav/American formulation reflecting feasibility within Yugoslavia;
3. Costs for residuals discharge reduction and modification processes be developed on the basis of existing Yugoslav costing methods and procurement policies (i.e., items stipulated by law to be manufactured in Yugoslavia be costed as such, import taxing policies and funding and credit arrangements, as defined by law, be followed);
4. The study of the Ljubljana area be limited to the five-commune area which represents the "crux" of the REQM problem in the region;
5. Where environmental modeling is not possible, percent reductions in residuals discharges be used as the environmental quality targets;
6. System variables assumed static and deterministic and final demand assumed fixed;
7. Locations and levels of activities assumed fixed;
8. The analysis to reflect 1975-1976 conditions.

Study Outputs

Rarely is it possible to investigate the effects of all variables on AEQ and on REQM costs. Therefore, specific analytical outputs must be delineated which are realistic, given the conditions for analysis. The following study outputs were defined:

1. Estimate least cost REQM strategies to achieve specific levels of environmental quality, where environmental quality is defined by a set of indicators relating either to ambient concentrations or to limitations on the discharge of various residuals and;

2. Demonstrate the effects of particular REQM measures on environmental quality and REQM costs.

How study outputs are defined directly affects both the analytical methodology used and the allocation and analytical resources. Therefore, a brief discussion of the rationale for the selected study outputs is merited. Yugoslavian society has been, and still is, undergoing rapid economic growth, with Slovenia being one of the fastest growing areas of the country. Although there is a clear concern with the increasing deterioration of some aspects of environmental quality which has accompanied that growth, there is also a concern that the physical measures to improve environmental quality will be "too" expensive. Thus, the first study output defined reflects the concern with finding the least-cost strategy to achieve any given level of environmental quality. The second study output defined reflects the view that, given limited resources in Slovenia to improve environmental quality, it would be helpful to have information on the effects and costs of particular single REQM measures which are under discussion and/or might be adopted. Both outputs are consistent with the goals stated earlier.

Specifying levels of environmental quality requires specifying the residuals to be analyzed. It is seldom possible, and in fact may neither be desirable nor necessary, to analyze all residuals. Not all residuals are likely to be important in a given region, or of equal importance. Therefore, some basis must be used to select the residuals to be analyzed. For the Ljubljana study the following five criteria were used:

1. For the residual there be either available data on generation or that institutional arrangements be available for collecting raw data on generation;

2. The residual be a generally accepted indicator of environmental quality;

3. There be "known" physical measures for reduction of the discharge of the residual to, or disposal in, the environment;

4. The residual be relatively important with respect to other residuals, in terms of residuals management costs and/or damages; and

5. The analysis of the number of residuals selected be feasible within the analytical resources available.

Table 5-1 shows the residuals selected for analysis and the level of quality specified, representing, where possible, Yugoslav environmental quality standards. The set shown is referred to as the "base" level of environmental quality. Other levels of environmental quality were also specified in order to illustrate various levels of quality which can be achieved at different costs.

Analysis for REQM

As indicated in Chapter 1, REQM involves the complete set of activities required to produce a desired level of AEQ, and an REQM management strategy consists of physical measures plus implementation incentives plus institutional arrangements. The Ljubljana study dealt primarily with that part of REQM involving the analysis of physical measures to improve environmental quality. Given the various residuals generating and discharging activities in the area, the analysis involved: categorizing the activities; estimating residuals generation by each category; developing individual (on-site) and collective

Table 5-1. Residuals Selected for Analysis and Base Set of Environmental Quality Targets

Environmental Quality Indicator	Abbreviation	Targets for "base" level of EQ: degree of discharge reduction; ambient concentration; quality of landfill operation
Liquid		
Five-Day Biochemical Oxygen Demand	BOD_5	80%
Total Suspended Solids	TSS	80%
Wastewater	WW	--
Gaseous		
Sulfur Dioxide	SO_2	150μg/m³ [a]
Total Suspended Particulates[b]	TSP	150μg/m³ [a]
Carbon Monoxide	CO	50%
Hydrocarbons	HC	50%
Nitrogen Oxides	NO_x	10%
Solid		
Mixed Solid Residuals (primary and secondary)[c]	MSR	Good quality sanitary landfill

[a]1972 Yugoslav standard; μg = micrograms.

Highest seven-day average during a four-month period.

[b]Total suspended particulates refer to the measurement of the ambient concentration of suspended particles of various sizes from various sources. The term particulate(s) refers to the discharge of particulate matter from exhaust stacks, some portion of which comprises the measured total suspended particulate concentration.

[c]Secondary residuals are those generated in the process of modifying primary residuals, i.e., sludge is a secondary residual produced from the modification of primary waste water in a sewage treatment plant.

(off-site) residuals discharge reduction cost functions; developing, where possible, environmental models for translating residuals discharges into the resulting impacts on AEQ; and finally, positing and evaluating sets of physical measures for reducing residuals discharges to the environment.

Analysis of Activities

Separating Activities into Activity Categories. After the environmental quality indicators were selected, activities in the Ljubljana study area were classified into activity categories with respect to economic function and considering the following four factors: (1) the relative importance of the residuals generating activity with respect to other generators of the same residual; (2) the relative importance of various variables affecting residuals generation associated with each activity; (3) the extent to which identifiable activity units can be affected by an REQM strategy; and (4) the available analytical resources--data, manpower, computational facilities, time. Seven activity categories were delineated: industrial, commercial, residential, institutional, transportation, power plant, and collective residuals handling and modification activities (municipal incinerators, municipal sewage treatment plants). These seven categories were, in turn, disaggregated into thirty subcategories. To illustrate the considerations in defining activity subcategories, the procedures used to disaggregate the residential and industrial activity categories are described.

Residential activities were divided into two subcategories: multi-flat residences, those comprised of more than two family dwelling units; and single-flat residences, those comprised of one or two family dwelling units. The rationale for this disaggregation is that these subcategories reflect significant differences in such variables as building structure, socioeconomic status, space, ownership, and heating technology, all of which affect residuals generation. For example, with respect to water use, in 1970 seventy percent of all flats in multi-flat residences had bath facilities, while only 59 percent of the flats in single-flat residences had bath facilities. Almost all single-flat residences have yard and garden areas which are watered, while multi-flat residences usually have none. These factors imply differences in water usage and hence in subsequent liquid residuals generation.[6]

With respect to generation of gaseous residuals, about three-fourths of the multi-flat residences have on-site large capacity energy conversion units installed and operated by housing maintenance companies. The remaining one-fourth is centrally heated from power plants. In contrast, single-flat residences generally have very small capacity energy conversion units of varying types of technology. Virtually all single-flat units are individually heated; almost no units are heated from central power plants. These factors imply differences in the generation of gaseous residuals and in applicable residuals discharge reduction measures.

Mixed solid residuals generation also differs between single-flat residences and multi-flat residences. Yard wastes and ash from solid fuel energy conversion processes can be significant components of residential mixed solid residuals generation. Depending upon the time of year, yard wastes and/or ash account for more than 50 percent of mixed solid residuals generation in single-flat residences. Because energy conversion technologies and the types and quantities of fuels used are different for single-flat residences and multi-flat residences, the corresponding ash components of mixed solid residuals generation differ.

These differences between multi-flat residences and single-flat residences in residuals generation are sufficient to warrant the disaggregation of residential activity, provided that REQM strategies and/or REQM costs can be differentiated between the two subcategories. Although additional data collection is required because of this disaggregation, more accurate delineation of residuals generation, residuals reduction measures, and costs is possible than if all residences were lumped together.

Delineation of subcategories of the industrial category was based on the Yugoslav standard industrial classification (SIC). This classification subdivides industrial activities primarily in relation to production processes and product mixes, and to a limited extent by the type(s) of raw materials used. These three variables, in addition to prices of factor inputs, are the major ones affecting residuals generation in industry. For the Ljubljana area study, ten industrial subcategories were delineated--nine for major SIC categories and one for "all other industries." Subcategories for the other five activities were defined in a similar manner.

Estimating Residuals Generation. For each subcategory some input or output "production" or activity parameter must be selected which indicates the level of activity, e.g., tons of product output, kilowatt hours of energy generated, barrels of crude petroleum processed, number of employees, number of inhabitants. The number of employees is usually readily available information and is frequently used in regional planning to specify present and to project future activity levels. However, for industrial activities, tons of product output or other units of physical output were used, because of the *direct* relationship between units of physical output and residuals generation.

Residuals generation coefficients are then expressed in terms of residuals generated per some unit of product output or activity level. For example, for an industrial activity mixed solid residuals generation is expressed in kilograms per metric ton of product output; BOD_5 generation in residences is expressed in grams per day per inhabitant; HC generation by automobiles is expressed in grams per vehicular kilometer traveled.

That all of the nonproduct outputs of materials and energy from an activity are not necessarily residuals should be emphasized. Some degree of materials and/or energy recovery may occur, depending on the relative value of the recovered materials and energy, with use of the recovered flows as inputs into either the same activity or another activity. For example, in developing solid residuals generation coefficients, recovered, salable non-product outputs were explicitly considered, so that such outputs were not included in the quantities of solid residuals estimated. Explicit consideration of recovery was not necessary for liquid and gaseous residuals, because either no recovery took place or recovery was internal to the activity.

Residuals generation coefficients can be developed in two ways: (1) by adopting and/or modifying coefficients found in the literature; (2) by deriving coefficients from empirical data collection and analysis. The coefficients used in the Ljubljana study resulted from the application of both procedures. For each subcategory the coefficient represents some average value; some individual operations in each activity subcategory have higher, some lower, generation coefficients.

In Yugoslavia there is presently little or no information, published or unpublished, on residuals generation or discharge. Consequently, much caution had to be used in adopting coefficients from non-Yugoslavian sources which reflect different cultural and economic conditions.[7] For example, a much-used

Western European coefficient for BOD_5 generation, developed in the Federal Republic of Germany, is 54 g/person/day, which is defined as one population equivalent (P.E.). However, sampling in the five-commune area produced a coefficient of 69 g/person/day--a substantial difference. So-called "foreign coefficients" were used only where they seemed to be reasonably representative of specific Yugoslav conditions. For example, U.S. coefficients were adopted for: the food and pulp and paper industries, for which production processes and product output mixes matched their U.S. counterparts; SO_2 and particulates for all activities where energy conversion technology could be matched with Yugoslav technologies; and generalized coefficients for CO, HC, and NO_x based on types of fuel combusted in all activities. For gaseous discharges from vehicular travel, West German coefficients for CO, HC, and NO_x were used, based on vehicle types similar to those in the five-commune area.

With respect to the empirical analyses, it was not possible to have sampling programs as rigorous and comprehensive as might be desired--that is, to obtain representative samples--even with the assistance of the local agencies. Considering resource constraints it was necessary to make judgments as to which activities were more important to sample than others. Consequently, no sampling was performed for BOD_5 and TSS discharges from institutional and commercial activities, because these activities represented only 1-2 percent of regional wastewater (WW) flow. On the other hand, intensive sampling of residences and industry was performed because they represent approximately 50 and 45 percent, respectively, of total regional wastewater flow (excluding discharge of cooling water from power plants). Residuals generation coefficients for solid (MSR), liquid (WW, BOD_5, TSS), and gaseous (SO_2, particulates, HC, CO, NO_x) residuals were developed for each activity.

Given the operating level for each activity and the estimated generation coefficients for the activity, residuals generation in the study area was estimated by simply multiplying the corresponding levels of production or other measure of level of activity by the relevant generation coefficients. Table 5-2 shows the total generation of residuals in the study area by activity category, for 1972. It is clear from table 5-2 which activities are the principal generators of each residual and hence for which activities residuals discharge reduction (RDR) cost functions should be constructed.

Developing Residuals Discharge Reduction (RDR) Cost Functions. Selecting RDR measures to construct RDR cost functions requires evaluation of factors such as the resources available within the society to install certain physical measures, technological differences in production processes, the availability of alternative technologies from foreign sources, possible barriers to the purchase of foreign technologies, and the evaluation of the extent to which a particular physical measure is perceived as a realistic option in the society. Available data, available analytical resources, and the perception problem limited the set of possible RDR measures investigated to relatively few physical measures. In the context of the Ljubljana study, it appeared doubtful that such measures as changing manufacturing production processes or family consumption patterns would be considered feasible, even if data on such possibilities could have been obtained. In addition, with the resources available it was not possible to develop RDR cost functions for each individual activity unit.

The basic premise used in constructing cost functions is that physical RDR measures can be incrementally applied to a specific activity category in a manner

Table 5-2. Residuals Generation in the Ljubljana Area, 1972[a]

Residual	*Industrial*	*Residential*	*Commercial*	*Institutional*	*Transportation*	*Power Plants*	*CRHM*[b]	*Totals*
Gaseous								
SO_2 10^3 Metric Tons	5.9	3.7	2.7	c	c	14.8	c	27.1
% of Total	21.8	13.7	10.0			54.6		100.0
Par. 10^3 Metric Tons	2.6	0.9	0.7	c	c	1.3	c	5.5
% of Total	47.6	16.4	12.8			23.6		100.0
CO 10^3 Metric Tons	0.2	1.9	1.6	c	22.2	0.1	c	26.1
% of Total	0.1	7.3	6.2		85.1	Neg.		100.0
HC 10^3 Metric Tons	0.1	0.4	0.3	c	0.8	0.1	c	1.7
% of Total	5.8	23.5	17.6		47.0	5.8		100.0
NO_x 10^3 Metric Tons	1.4	0.7	0.3	c	0.7	5.1	c	8.2
% of Total	17.1	8.5	3.7		8.5	62.2		100.0
Liquid								
BOD_5 10^3 Metric Tons	82.7	63.4	6.6	27.7	0.5	0.1	Neg.	181.5
% of Total	45.6	34.9	3.6	15.3	0.6	Neg.	Neg.	100.0
TSS 10^3 Metric Tons	32.9	57.3	5.9	25.3	0.4	0.1	Neg.	122.3
% of Total	26.9	46.9	5.0	20.9	0.3	Neg.	Neg.	100.0
WW 10^3 Metric Tons	18.1	11.4	2.1	2.4	0.9	52.7[d]	Neg.	88.2
% of Total	20.6	12.9	2.5	2.7	1.0	59.8[d]	Neg.	100.0
Solid								
MSR 10^3 Metric Tons	55.0	40.7	8.3	5.2	3.6	60.5	0.1	174.0
% of Total	32.0	23.3	4.8	3.0	2.2	34.6	Neg.	100.0

Neg. = Negligible; Par. = particulates.

a All values are rounded.

b Collective residuals handling and modification activities.

c Included in totals for commercial category.

d Primarily represents discharges of cooling waters.

such that the costs incurred in each increment can be calculated and that there will be a corresponding calculable reduction in residuals discharged from that activity. For example, installing cyclonic filters of a particular design on all boilers within a specific range of capacity in an industrial category results in so many tons of particulates no longer discharged to the atmosphere. The objective in developing RDR cost functions is to find the least cost method of achieving each additional increment of discharge reduction, including the costs of disposing of secondary residuals generated by each increment, such as the particulates collected in a filter.

Each cost function is a graphic or tabular representation of the costs required to reduce the discharge of a residual by various amounts, costs which are in addition to the normal costs of production and which would not be incurred in the absence of constraints on discharges of residuals ("pollution controls"). It should be emphasized that in some cases the measures installed result in recovery of some usable materials and/or energy, or otherwise reduce operating costs. Whatever savings are achieved must be subtracted from the gross costs, in order to obtain the net costs of RDR. If the measures increase other operating costs, the additional costs must be included.

The costs considered for each incremental application of an RDR measure to an activity include initial capital construction costs--new equipment, site preparation, installation, and annual operating and maintenance (O&M) costs--fuel, electrical energy, labor, maintenance materials. Also included are the capital costs of replacing equipment which has a physical life less than the time period of analysis. All costs subsequently indicated are in 1974 New Dinars (N.D.).

Table 5-3 indicates the physical measures and the residuals which they affect. These physical measures comprised the population from which cost functions were developed. Note that table 5-3 also indicates whether the measure is applied on-site or off-site. Off-site application of a physical measure takes place in a collective residuals handling and modification facility and involves the collection of residuals from two or more individual activities for modification and/or disposal into the environment, i.e., a municipal sewage treatement plant. On-site application of a residuals discharge reduction measure takes place, as the term infers, at the site of residuals generation from an individual activity, i.e., substituting light liquid fuel for coal by modifying individual boilers at activity locations. The procedure for developing an RDR cost function for a collective facility is essentially the same as developing an on-site function.

The development of an on-site RDR cost function for reducing SO_2 discharges from the metal processing industry is used to illustrate the procedure. Three measures to reduce SO_2 discharges in the industry were analyzed: (1) conversion of existing coal-using boilers to light liquid fuel boilers; (2) conversion of existing heavy liquid fuel-using boilers to light liquid fuel boilers; and (3) desulphurization of the heavy liquid fuel for use in the existing heavy liquid fuel-using boilers. For analysis of measures (1) and (2), the following information was obtained on all boilers used in the metal processing industry: boiler technology; current fuel type; capacity; and age of boiler. Based on this information Yugoslav engineers estimated the capital costs--equipment purchase and installation--for modifying/installing burners and storage tanks. The cost differential in using the higher priced liquid fuel was calculated by estimating energy conversion efficiencies for each boiler,

Table 5-3. Physical Measures for Reducing Residuals Discharges Analyzed in Ljubljana REQM Study

REQM Physical Measure and Extent of Modification	Type of Residual Affected							
	Gaseous					Liquid		Solid
	SO_2	Part.	CO	HC	NO_x	BOD_5	TSS	MSR & Sludge
Cyclonic filter to remove 95% of particulates in stack gas (on-site)		X						0
Substitution of light liquid fuel with 1% sulfur, essentially no ash, for coal with 2% sulfur, 12.5% ash (on-site)		X	X	X	X			a
Substitution of light liquid fuel with 1% sulfur, essentially no ash, for heavy liquid fuel with 3% sulfur, essentially no ash (on-site)	X	X	X	X	X			
Desulphurization of heavy liquid fuel (to 0.5% sulphur)[b] (off-site)	X	X						
Off-site central heating from coal-fired power plants to Residential, Commercial, and Institutional activities in feasible connection area	X[c]	X	X[c]	X[c]	X[c]			0
Park and Ride, to reduce daily vehicle kilometers traveled in five-commune area			X	X	X			
Idling speed adjustment, applied twice a year to all gasoline-fueled vehicles			X	X	X[c]			
Sedimentation/activated sludge plant, collective facility (off-site)						X	X	0
Sedimentation/aerated stabilization plant (on-site)						X	X	0
Sanitary landfill, collective facility (off-site)								X

X Indicates measure applied to indicated residual.
0 Indicates application of measure results in generation of secondary residual of type indicated.
[a]Decreases the quantity of ash to be handled in the study area.
[b]Desulphurization is assumed to be done outside the study area.
[c]Results in net increase of discharge.

based on boiler technology and age, by subtracting the cost of usable energy from coal fuel or heavy liquid fuel currently used from the cost of an equivalent amount of usable energy from the light liquid fuel. The additional cost of system maintenance was estimated to be negligible. For measure (3), it was assumed that a desulphurization plant would be constructed outside of the study area to provide desulphurized fuel for use not only in the Ljubljana area but in other areas of the country as well. The average cost of the desulphurized fuel produced by such a plant was estimated by Yugoslav engineers and was used as the unit price for the amount of heavy liquid fuel necessary to generate an equivalent amount of energy in the relevant boilers.

Reductions in SO_2 discharges were estimated by calculating the difference between SO_2 discharges under present conditions and SO_2 discharges with one or more of the alternative measures in place. The capital and O&M costs and the corresponding reduction in SO_2 discharges for each alternative applied to the metal processing industry are shown in table 5-4A. Table 5-4B shows the capital and O&M costs and reduction in SO_2 discharge for various combinations of these alternatives. Similar costs for various alternatives were developed for three other industrial subcategories--pulp and paper industry, chemical industry, and all other industries.

It should be emphasized that each physical measure applied to reduce SO_2 discharge also affects, to varying degrees, the discharges of other gaseous residuals from these activities--particulates, CO, HC, and NO_x. For example, converting existing coal-using boilers to light liquid fuel boilers in the metal processing industry to reduce SO_2 discharges results in a 74-percent reduction of SO_2 discharges and simultaneous reductions in particulates, CO, HC, and NO_x of about 80, 99, 84 and 74 percent, respectively.

Table 5-4A. Residuals Discharge Reduction Costs[a] and Percent SO_2 Discharge Reduction Achieved for Each Alternative Applied to the Metal Processing Industry

	Alternative	Capital Costs 10^6 N.D.	O&M Costs 10^6 N.D./yr.	Percent Reduction in Discharge of SO_2
I	Convert boilers using coal[b] to light liquid fuel units	2.7	3.2	74.0
II	Desulphurize heavy liquid fuel for boilers using heavy liquid fuels	None[c]	0.5	5.6
III	Convert boilers using heavy liquid fuels to light liquid units	0.2	0.5	4.5

a All costs are rounded and in 1974 N.D.
b Includes all solid fuels.
c The capital costs of a desulphurization plant are not shown as a capital cost for the metal processing industry. They are reflected in higher fuel prices paid by the industry and therefore are included in O&M costs.

Table 5-4B. SO_2 Residuals Discharge Reduction Costs[a] and Percent SO_2 Discharge Reduction Achieved for Joint Application of Alternatives to the Metal Processing Industry

Combined Alternatives	Capital Costs 10^6 N.D.	O&M Costs 10^6 N.D./yr.	Percent Reduction in Discharge of SO_2
I only	2.7	3.2	74.0
I plus II	2.7	3.7	79.6[b]
I plus III	2.9	3.7	78.5[b]

a All costs are rounded and in 1974 N.D.
b For the region, percent reductions for applying alternatives I plus II and I plus III to the metal processing industry would be about 2.2% and 2.1%, respectively.

Environmental Modelling

Environmental models describe the impacts on AEQ of residuals discharged into the environment. Such models must be formulated in terms of the quantities of specific types of residuals discharged at specific locations of activities in a given area.

In the Ljubljana study, given the resources available, the following decisions were made:

(1) Because: deteriorating air quality was the most serious AEQ problem in the area; air quality was a problem for which little analysis had been done; and no empirical data were available with respect to discharges of gaseous residuals and limited data were available with respect to ambient air quality, it was concluded that relatively simple modelling of air quality would provide useful information for management decisions at present.

(2) Because the lack of water quality data in the area precluded even simple "first-cut" modelling of water quality, an attempt would be made only to delineate the dimensions of the water quality problem and discuss the implications for water quality modelling in the study area, and in the larger natural watershed of which the study area is a part.

(3) No modelling of the groundwater system would be attempted.

Analyzing Ambient Air Quality. In analyzing air quality, the objective was to construct a model which would enable the evaluation of various physical measures for improving air quality in terms of their costs and impacts on ambient air quality. To determine what type of model to use, the physical and meteorological characteristics of the area were first examined. Review of the available data showed that the five-commune area included most of the air quality problems in the Ljubljana Basin, on the basis that approximately 90 percent of all SO_2 and particulate discharges in the basin presently occur within the five-commune boundary. Also, the five-commune area includes 95 percent of the population and approximately 90 percent of the land area in the basin. Examination of the meteorological records showed that: (1) there was little variation in mean meteorological conditions, such as wind speed, wind direction, and precipitation, throughout the basin; and (2) little or no ventilation[8] occurs in the basin in the winter months when air quality is the poorest.

The above justified the following simplifying assumptions.

(1) Discharges of gaseous residuals from sources outside the five-commune area contribute insignificantly to ground level concentrations within the five-commune area, and therefore do not have to be analyzed.

(2) Discharges of gaseous residuals within the study area result in negligible adverse impacts outside of the study area.

(3) Because relatively complete mixing occurs in the basin, sufficently accurate mean daily ground level concentrations in the area of SO_2 and TSP for management decisions can be obtained by averaging the daily average measurements at the eighteen air quality monitoring stations in the five-commune area.

Given the above assumptions and the fact that ambient air quality is a function of the quantity of gaseous residuals discharged and of meteorological conditions, a relatively simple air quality model--a multiple linear regression model (MLRM) was applied. The model was of the following form:

$$Y_S \text{ or } Y_{TSP} = B_1 + B_2 X_1 + B_3 X_2 + \ldots B_k X_n + E,$$

where, Y_S or Y_{TSP} = average ambient concentration of SO_2 and TSP, respectively;

$X_1 \ldots X_n$ = a set of meteorological and residuals discharge variables;
$B_1 \ldots B_k$ = set of coefficients to be estimated; and
E = random disturbances.

The independent variables for use in the regression analysis were of two types: environmental variables, those which represent meteorological conditions and represent the assimilative capacity of the atmosphere; and residuals discharge variables, which reflect discharges from various activities and can be affected by REQM strategies. The four environmental variables are:

X_1 -- precipitation per day in mm, averaged over seven-day periods;

X_2 -- average wind speed per day in m/sec, averaged over seven-day periods;

X_3 -- meteorological stability per day--five ordinate classes averaged over seven-day periods: (1) markedly cyclonic, (2) cyclonic, (3) undefinable, (4) anticyclonic, (5) markedly anticyclonic; and

X_4 -- duration of inversion per day--four classes averaged over seven-day periods; (0) 0 hours, (1) $0 < X_4 \leq 8$ hrs, (2) 8 hrs $< X_4 \leq 16$ hrs, (3) 16 hrs $< X_4 \leq$ 24 hrs.

The residuals discharge variables are activity groupings based on types and amounts of fuel used. The eight residuals discharge variables are:

X_5 -- Chemical Industries, S.I.C. 120;

X_6 -- Metal Processing Industries, S.I.C. 117;

X_7 -- Pulp & Paper Industries, S.I.C. 123;

X_8 -- All Other Industries;

X_9 -- Power Plants;

X_{10}-- Single-flat Residences;

X_{11}-- Multi-flat Residences, and

X_{12}-- All Other Activities.

Daily air quality measurements were available for 1970, 1971, and 1972 from the eighteen monitoring stations in the five-commune area. Using these data, mean monthly concentrations of SO_2 and TSP were computed for each month in the three-year period. The means for the winter, spring/fall, and summer climatic periods--each period being four months in duration--were sufficiently different so that it was decided to apply the MLRM separately to each of the periods. However, only the SO_2 and TSP equations for the winter climatic period were used for estimating the effects on ambient air quality of various physical measures, because only during this period does the average ambient SO_2 concentration exceed the current SO_2 standard. The average for 1970-1972 of ambient concentrations of TSP during the winter season is the highest of the three seasons, but is still slightly lower than the current ambient TSP standard. However, because the average was not much below the current standard, and because both population and production in the area have increased since 1972, analysis of TSP was considered useful.

In order to have sufficient observations for a statistically meaningful sample, twelve months of daily meteorological data for each of the three climatic periods were divided into seven-day periods, and the means for those periods calculated. This yielded 52 observations of the dependent variable for each of the climatic periods.

The regression equation for each climatic period is generated by using a "backward stepwise" least squares regression routine[9] with a principal components subroutine, to regress the twelve independent variables described above on the dependent variable Y. First, the principal components subroutine is used to form artificial variables. This must be done to reduce multicollinearity between variables that are interrelated. These artificial variables are then regressed on Y using the backward stepwise regression routine. The program continues through iterations of the regression, removing the least significant artificial variable each time, until all variables are significant at a level of less than or equal to 0.1 (t-test). The remaining artificial variables are then translated to give the explanatory power of the original twelve variables.

The model is used in the evaluation of physical measures for improving ambient air quality by inserting in the equations the changes in the quantities of residuals discharges that would result from the application of those measures and calculating the estimated resulting ambient concentration. By repeating the procedure, the costs of achieving various levels of ambient SO_2 and TSP concentrations are estimated.

Table 5-5 shows the coefficients for the two-winter period equations and the results of the statistical tests applied to the intermediate principal components regression equations. The statistical tests used were: coefficient of determination for the equation (R^2); F test for the whole equation; the Durbin-Watson statistic for serial (or auto-) correlation; and "Student's t" test for the significance of individual coefficients. On the basis of these tests, sufficient confidence can be placed in the results of the winter period regression equations for SO_2 and TSP so that these equations can be used to estimate the effects of REQM measures. But it should be emphasized that no statistical analysis is meaningful unless there is a corresponding physical relationship which is consistent with the statistical results.[10] Careful interpretation is an essential ingredient in the use of statistical models.

Implications for Analyzing Ambient Water Quality.[11] Mathematically expressed models of river water quality and ground water quality would contribute to decision-making in the region, both by improving knowledge of existing stream and aquifer behavior and by aiding in the estimation of changes in water quality resulting from alterations in stream flow, groundwater withdrawals, and discharges of liquid residuals. Such models are essential tools in the evaluation of alternative water quality and water supply management strategies. Although for the REQM study of the five-commune area environmental models of the stream and of the ground-water aquifer would be useful, existing data were insufficient to develop such models. However, based on available knowledge

Table 5-5. Results of Multiple Linear Regression Model Application for SO_2 and TSP

	SO_2	TSP
	Winter	Winter
Statistic Relating to Regressions of the 12 Artifical Variables		
• number of significant artificial variables (t-test level of significance ≤ 0.1)	6	5
• R^2	0.73	0.61
• F-test (Level of significance ≈ 0.00)	24.5	14.4
• Durbin-Watson test (1% level of significance)	No A.C.[a]	Incon-clusive
• % of Y variance explained in regression	78%	89%
Translated Values of Coefficients of Original Independent Variables (X)		
X_1	-.10	-.00[b]
X_2	-.24	-.31
X_3	-.13	+.11
X_4	+.29	+.37
X_5	+.17	+.13
X_6	+.07	+.09
X_7	+.07	+.17
X_8	-.10	+.02
X_9	-.11	+.00[c]
X_{10}	+.19	+.05
X_{11}	+.18	+.06
X_{12}	+.20	+.05
Constant	+.32	+.15

[a] A.C. = autocorrelation
[b] Slightly less than zero
[c] Slightly greater than zero

of the water quality and supply problems in the five-commune area, implications may be drawn concerning the extent and types of water quality modelling that might usefully be undertaken in the future. There are three major implications.

First, there is no apparent need to develop a water quality model of the Ljubljanica River. Because the Ljubljanica currently has such poor water quality, based on several indicators--as a result of receiving most of the liquid residuals discharges from the urban area--little insight into the investment decisions needed to improve ambient water quality would be gained by investing scarce resources to model it. Rather, efforts should be expended to obtain estimates of the total quantities of organic materials and suspended sediment discharged into the Ljubljanica from different point and nonpoint sources, and of the costs to reduce such discharges. It is likely that such information would indicate where resources should be invested first to begin to improve water quality in the Ljubljanica.

Second, because reducing liquid residual discharges in the five-commune area would have little or no effect on water quality of that segment of the Sava River in the five-commune area, modelling only that segment of the Sava would be of little value at present to decision makers within the five-commune area for two reasons. One, the quality of the Sava River in the five-commune area can only be affected by reduction in residuals discharges upstream from Ljubljana, over which discharges the decision makers in the five-commune area have no control. Two, liquid residuals discharges in the five-commune area do have significant impacts on ambient water quality downstream at Zagreb and below in the Republic of Croatia,[12] but at present there are no ambient water quality standards which have to be met at the study area boundary.

If water quality modelling of the Sava River is to contribute to REQM decisions in the Ljubljana area, the modelling must include at least the entire Upper Sava River Basin. The limited data presently available on ambient water quality and residuals discharges in that basin did not justify attempting to develop a model of that segment of the river. The modelling would have to include much of the Sava downstream from the Ljubljana area as well, particularly if constraints on water quality of the Sava at the boundary of the five-commune area were to be analyzed.

Third, with respect to improving the capacity to manage the aquifer system, a model of flow into, within, and from the aquifer will soon be required. Among the questions that must be answered concerning the aquifer are those relating to the effects of the quantity and location of pumping from the aquifer and/or recharging it on the yield and water quality of the aquifer. Because of the limited information available concerning the Ljubljana aquifer system, it is not possible to recommend a specific modelling approach at this time, only that an attempt to model the aquifer should be given high priority, because of serious uncertainties about its productive capacity and water quality in the near future, given present trends.

Based on the reasons discussed above, it was decided that no water quality models would be attempted for the REQM study of the Ljubljana area. Therefore, the effects of physical measures for managing liquid residuals were developed in terms of the degree of reduction in the discharge of those residuals.

Developing and Analyzing Sets of Physical Measures for Reducing Residuals Discharges and Improving AEQ

A set of physical measures can be comprised of: a single physical measure directed at one residual from a single activity; various physical measures directed at one residual generated by multiple activities; various physical measures directed at several residuals generated by one or more activities. The number of sets of physical measures developed in an analysis for REQM is a function of: (1) the number of different residuals discharge reduction measures developed for each individual activity; (2) the number of activities; (3) the resources available for analyzing sets of physical measures; and (4) the number of alternative levels of environmental quality investigated. In the Ljubljana study, a base set of environmental quality targets was specified, as indicated in table 5-1. Four other sets of environmental quality targets were also analyzed, in order to show the costs of achieving different levels of quality and to illustrate the trade-offs among residuals and environmental media.

Table 5-3 listed the physical measures which can be applied to each of the activities for modification of residuals, and the residuals to which they apply. Some of these measures are mutually exclusive, i.e., they cannot be applied simultaneously to the same activity. For example, the alternative of desulphurization of heavy liquid fuel cannot be applied in addition to applying the alternative of shifting from heavy liquid fuel to light liquid fuel. Similarly, the effluent from an on-site sedimentation and aerated stabilization plant is not likely to be discharged into a collective activated sludge facility. Because of limited analytical resources, each physical measure was applied at only one level of reduction. For example, cyclonic filters were designed to remove 95 percent of the particulates from the discharge stream.

Sets of physical measures were developed from the individual physical measures listed in table 5-3. The universe of possible alternative physical measures is comprised of ten physical measures which can be applied in various combinations to eleven activity categories and subcategories--metal processing industries, chemical industries, pulp and paper industries, all other industries, power plants, transportation, commercial, institutional, and collective residuals handling and modification activities. Not all physical measures are relevant to all activities, because not all activity categories are significant generators of each of the residuals of interest. Each physical measure has a physical service life ranging from six to forty years, depending on the characteristics of the material items included, at the end of which the salvage value of material items is assumed to be zero. The same type of physical measure can have different service lives for different activity categories.

The first step in the development of sets of physical measures was to estimate the effects of the application of each physical measure to each activity category for which it was relevant, in terms of the amount of reduction in residuals discharge achieved and, where applicable, the improvement in AEQ achieved.

The second step was to calculate the present value of the time stream of capital and O&M costs, including replacement costs, of each physical measure applied to each activity category for the the forty-year period of analysis. The costs also include the capital and O&M costs of handling and disposing of the secondary residuals generated in the modification of primary residuals, such as disposal of fly ash collected in cyclonic filters.

Economic analysis[13] for decision-making for REQM strategies, as for all investment decisions, is complicated by inflation. As long as inflation is general, i.e., the same for all sectors of the enonomy, the evaluation of alternatives is not affected in the sense that their <u>relative</u> ranking will be unchanged. However, even if differential rates of inflation were to occur, it is virtually impossible to predict those rates, especially in a rapidly developing economy such as that of Yugoslavia. Inflation is also one of several factors which affects the choice of the social rate of discount to use in discounting the time streams of costs to their present values. A rate of 11 percent was used, as recommended by Yugoslav experts.[14]

Having calculated the present value of costs for each physical measure applied to each relevant activity in relation to each residual, the third step was to calculate the cost effectiveness of each measure applied, i.e., the amount of environmental quality achieved by a given expenditure. This was done in terms of percent reduction in ambient SO_2 or TSP concentration per 10^6 New Dinars of expenditure, or percent reduction in discharge of BOD_5, TSS, HC, NO_x, and CO, per 10^6 New Dinars of expenditure. For estimating the actual cost effectiveness of physical measures, this method is crude at best, because a given physical measure applied to an activity category will often reduce the discharge of more than one residual. For example, the Park and Ride measure will simultaneously affect discharges of HC, CO, and NO_x. Nevertheless, it is an operational procedure for initial selection of physical measures to comprise sets.

The cost effectiveness of each physical measure-activity category combination suggests the sequence of addition of such combinations to find the least-cost set of physical measures to meet the specified targets. The procedure is to begin with the most cost effective physical measure-activity combination and add combinations in order of effectiveness until all E.Q. target values are met or exceeded, or all combinations are added. In the Ljubljana study, each physical measure-activity category was added in its entirety. That is, each physical measure was applied to <u>all</u> individual units in the activity category.

If more analytical resources and/or data had been available, it would have been possible to apply different RDR measures to subdivisions of activity categories. By limiting the options to activity categories as a whole, target values cannot be met exactly.

The physical measure-activity category combinations comprising the least cost set of physical measures, denoted SPM I, are listed in table 5-6. The E.Q. targets, the E.Q. levels actually achieved, and the associated present value of costs for the minimum cost set of physical measures are shown in table 5-7. The physical measures employed result in: (1) not achieving the HC, CO, and NO_x targets; (2) considerably exceeding the TSP target; (3) moderately exceeding the SO_2 target; (4) not achieving the BOD_5 target by a small amount; and (5) just achieving the TSS target. The total cost of SPM I is about 3.4 billion N.D.

Because the last increments of physical measures added to reduce TSS, BOD_5 and SO_2 discharges are relatively expensive, four additional sets of physical measures (II, III, IV, and V) were developed to illustrate the costs of achieving different levels of environmental quality. The costs of, and the environmental quality levels achieved by, these sets of physical measures are shown in table 5-8, along with the corresponding results for SPM I. For all sets of physical measures, all mixed solid residuals generated are deposited in good quality sanitary landfill and the same degrees of discharge reduction are achieved for CO, HC, and NO_x. For SPM II, the reductions in TSS and BOD_5 discharges are slightly less than for SPM I, a result accomplished by eliminating the last two increments of RDR measures for these two residuals. The ambient concentrations of SO_2 and TSP achieved are the same as for SPM I. SPM II costs about 2.4 billion N.D., or about 70 percent of the costs for SPM I, with only slightly lower reductions in TSS and BOD_5 discharges.

Table 5-6. Physical Measure-Activity Category Combinations Comprising The Least Cost SPM to Achieve Environmental Quality Targets

Physical Measure-Activity Category Combination	Environmental Quality Indicator	Degree of Discharge Reduction (DR) or Ambient Improvement (AI) In Region	Present Value of Capital Costs 10^6 N.D.	Present Value of O & M Costs 10^6 N.D.	Total Present Value of Costs 10^6 N.D.
Substitute Light Liquid Fuel (1%S) for Coal:					
-Chemical Industry	SO_2	7% AI	4	18	22
	TSP	8% AI			
-Pulp and Paper Industry	SO_2	19% AI	5	95	100
	TSP	30% AI			
-Metal Processing Industry	SO_2	6% AI	3	28	31
	TSP	10% AI			
-Multiple-Flat Residences	SO_2	11% AI	6	100	106
	TSP	4% AI			
-Total Effect of Measure-Activity Category on Other Gaseous Residuals	CO	5% DR	Included in each category above		
	HC	15% DR			
	NO_x	5% DR			
Change to Desulphurized Heavy Liquid Fuel (0.5%S):					
-Chemical Industry	SO_2	7% AI	b	25	25
-Pulp & Paper Industry	SO_2	8% AI	b	58	58
-Single-Flat Residences	SO_2	1% AI	b	13	13
Park/Ride Level II:					
-13% Reduction in VKT in Region (5-Commune Area)	CO	16% DR	300	480	780
	HC	9% DR			
	NO_x	1% DR			
Carburetor Idling Adjustment:					
-All Gasoline-Fueled Vehicles In Region	CO	17% DR	2	106	108
	HC	9% DR			
	NO_x	-- DR[c]			
On-Site Modification of Liquid Residuals for 12 Outlying Industrial Plants[d]/	TSS	7% DR	43	18	61
	BOD_5	23% DR			
Primary Settling and Activated Sludge for Collective Waste-water Treatment Plant	TSS	73% DR	1732	373	2105
	BOD_5	53% DR			
Sanitary Landfill	All MSR	--	15	11	26

a All values rounded; costs are in 1974 N.D.

b Capital costs are reflected in operating and maintenance costs for the consumer.

c Carburetor idling adjustment increases NO_x discharges while CO and HC discharges are decreased. The amount of increase in NO_x discharge could not be estimated at this time.

d Primary sedimentation at six plants, and primary sedimentation and aerated stabilization plant at six plants.

Table 5-7. Environmental Quality Targets, Environmental Quality Levels Achieved, and Present Value of Costs [a] for SPM I

Environmental Indicators	Target	Level Achieved	P.V. of Costs,[b] 10^6 New Dinars
SO_2	150 μg/m^3	135 μg/m^3	360
TSP	150 μg/m^3	75 μg/m^3	
BOD_5	80% reduction	76% reduction	2170
TSS	80% reduction	80% reduction	
CO	50% reduction	35% reduction	890
HC	50% reduction	25% reduction	
NO_x	10% reduction	6% reduction	
Mixed Solid Residuals (MSR)	Good quality sanitary landfill, all MSR	Good quality sanitary landfill, all MSR	30
		Present Value of Total Cost	3450

[a]All values rounded; costs are in 1974 N.D.

[b]40 years; 11% discount rate.

Table 5-8. Environmental Quality Levels Achieved and Corresponding Present Value of Costs for Various Sets of Physical Measures

Environmental Quality Indicator	SPM I		SPM II		SPM III		SPM IV		SPM V	
	E.Q. Level Achieved	P.V. of Costs, 10^6 N.D.	E.Q. Level Achieved	P.V. of Costs, 10^6 N.D.	E.Q. Level Achieved	P.V. of Costs, 10^6 N.D.	E.Q. Level Achieved	P.V. of Costs, 10^6 N.D.	E.Q. Level Achieved	P.V. of Costs, 10^6 N.D.
SO_2	135 μg/m^3	360	135 μg/m^3	360	170 μg/m^3	250	200 μg/m^3	180	110 μg/m^3	470
TSP	75 μg/m^3		75 μg/m^3		80 μg/m^3		80 μg/m^3		70 μg/m^3	
CO	35% DR	890	35% DR	890	35% DR	890	35% DR	890	35% DR	890
HC	25% DR		25% DR		25% DR		25% DR		25% DR	
NO_x	6% DR		6% DR		6% DR		6% DR		6% DR	
TSS	80% DR	2170	76% DR	1180	74% DR	840	63% DR	420	76% DR	1180
BOD_5	76% DR		71% DR		69% DR		61% DR		71% DR	
MSR	Good quality sanitary landfill, all MSR	30	Good quality sanitary landfill, all MSR	30	Good quality sanitary landfill, all MSR	30	Good quality sanitary landfill, all MSR	30	Good quality sanitary landfill, all MSR	30
TOTAL COSTS		3450		2460		2010		1520		2570

[a] All values rounded; costs are in 1974 N.D., 40-year time period, 11% discount rate.

For SPM III and SPM IV, the ambient concentrations of SO_2 and TSP achieved are higher than, and the reductions in TSS and BOD_5 discharges achieved are less than, those achieved by either SPM I or SPM II. In SPM III the last three increments of RDR measures to reduce BOD_5 and TSS discharges, and the last RDR increment used in SPM I to reduce SO_2 discharges are not included. For SPM IV, the ambient concentration of SO_2 achieved is higher than, and the reductions in TSS and BOD_5 discharges achieved are less than, the corresponding levels for SPM III; the ambient concentration of TSP is the same for both SPM III and SPM IV. These decreases in environmental quality are reflected in decreases in costs, the costs of SPM III and SPM IV being about 2.0 billion N.D. and 1.5 billion N.D., respectively, corresponding to about 60 percent and 45 percent of the costs for SPM I, respectively.

SPM V was developed because of the concern in the five-commune area for improving ambient air quality. For SPM V, the ambient concentrations of SO_2 and TSP achieved are lower than for SPM I and the reductions in discharges of TSS and BOD_5 are slightly less than for SPM I. The costs for SPM V are about 2.6 billion N.D., about 75 percent of those for SPM I.

Although each set of physical measures must be integrated with implementation incentives and institutional arrangements to form an REQM strategy, the results of the analyses of physical measures above have implications with respect to REQM in the Ljubljana area. The order in which these implications are discussed does not imply the order of their importance.

First, each increment of reduction in TSS and BOD_5 discharges above 70-80 percent by means of end-of-pipe measures, i.e., standard physical-biological sewage treatment, requires rapidly increasing incremental costs. Even if the benefits from greater than 70-80 percent reduction in discharges were high, efficient utilization of scarce resources suggests that REQM efforts should be oriented toward other physical measures to reduce TSS/BOD_5 generation, i.e., in-plant and/or in-household changes, rather than toward end-of-pipe modification. This orientation will become even more important as the economy of the area grows, and more demands are made on the finite environmental resources.

Second, the relatively low cost of good quality sanitary landfilling suggests that this physical measure should be a part of any REQM strategy for the area. The benefits in terms of precluding adverse impacts on groundwater quality, visual appearance, and public health are likely to be greater than the costs.

Third, the two physical measures for reducing CO, HC, and NO_x discharges from vehicular movements achieve only a relatively small decrease in such discharges. This suggests that major changes in the transport system within the five-commune area will be necessary if any significant improvement in these indicators of environmental quality is to be achieved. Major changes in motive power of individual vehicles and a major shift to mass transportation (with changed motive power) will be necessary.

Fourth, shifting to off-site supply for central heating can have negative impacts on ambient SO_2 concentrations. Three factors account for this. One, the shift to off-site supply results in an increase in the total amount of energy conversion required for heating purposes, because off-site supply induces the heating of more space in dwelling units. Two, high sulphur content solid fuel is used in power plants in contrast to liquid fuel and low sulphur content brown coal, which are presently the principl fuels used in on-site heating installations. The quantity of fuel used by the power plants is about half the total fuel combusted in all activities other than transportation. To shift from light liquid fuel or to shift to desulphurized heavy fuel at the power plants would involve major expenditures, given the present sources and qualities of fuel and large existing investments in coal combustion technology.[15] Three, wet-scrubbing and limestone injections to reduce SO_2 discharges from power plants have not been considered as feasible RDR alternatives because of excessive costs and technical difficulties in obtaining the relevant equipment in Yugoslavia. (Shifting to off-site supply for central heating does reduce particulate discharges, but it is not the least cost measure for reducing such discharges).

Fifth, the seasonal averaging period, winter, used in the analysis of ambient air quality masks the effects of short periods of several days duration, which occur quite frequently in the area. Even though the monthly mean ambient concentrations of SO_2 and TSP meet the respective standards, concentrations substantially higher than the standards can occur for short periods. In addition, analysis of historical data showed that in some months the monthly mean concentrations of SO_2 and TSP were 25 percent greater than the standards because of a few days of high concentrations (episodic conditions), rather than concentrations continually above the standard throughout the month.

The physical measures investigated in developing the sets of physical measures will effectively reduce ambient concentrations on the average. These measures, however, will not enable achieving the ambient SO_2 standard during the short-period, critical conditions. The only physical measure which can be applied intermittently is the substitution of desulphurized heavy liquid fuel for the high sulphur heavy liquid fuels. If such action is insufficient, the actual level of activity during critical periods must be reduced.

Sixth, some of the physical measures which can be applied to a given activity to reduce the discharge of one (or two) residuals are mutually exclusive or counterproductive. With respect to the former, either desulphurizing heavy liquid fuel or changing from heavy liquid to light liquid fuel can be applied at a given activity, but not both. With respect to the latter, for certain activities the substitution of light liquid fuel for coal to reduce the discharge of SO_2 would preclude installing cyclonic filters which comprise the most cost-effective measure for reducing particulate discharges from those activities.

REQM Strategies

The discussion in the preceding section focused on the first element of REQM strategies--physical measures for reducing residuals discharges. In this section relationships between those physical measures and the other two elements of REQM strategies--implementation incentives and institutional arrangements--are discussed in the Yugoslav context.

Selecting Physical Measures

The analysis of physical measures in the Ljubljana study separately from the other two elements of REQM strategies does not mean that the physical measures analyzed were chosen without consideration of implementation incentives and institutional arrangements. In accord with the goal enunciated in the beginning--that is, to make the analysis as Yugoslav-oriented as possible--consideration of the feasibility of implementing any given physical measure was a factor in selecting the measures to be analyzed. This meant identifying the principal actors, both organizations and individuals, who could provide the necessary information and assist in selecting and developing physical measures for analysis. For example, selecting physical measures for improving air quality, particularly with respect to SO_2 and TSP, received the most scrutiny for three reasons. One, the air quality problem is perceived by many people to be the most critical environmental quality problem in the study area. Two, much governmental interest, principally the Slovene Republic Committee for Air Quality, has been in improving air quality. Three, it was possible to develop an air quality model as described earlier. The Republic Committee assisted in formulating physical measures for reducing discharges of SO_2 and particulates; other Yugoslav sources suggested measures for reducing discharges of CO, HC, and NO_x.

To select feasible measures for reducing SO_2 and particulates discharges, two sessions were held with the Committee. In the first, previous work on the air quality model for Ljubljana and published material were presented to the Committee and other interested parties. The second was a closed working session to define physical measures which were considered feasible to implement. In the working session a questionnaire containing a list of potential physical measures was presented to the Committee. After much discussion three measures were selected: fuel substitutions; central heating from power plants; and the use of stack-gas cleaning equipment. The specific details for each physical measure were worked out by the project staff. Similar procedures were used to select the other physical measures analyzed in the study.

Implementation Incentives

Implementation incentives are the means by which the residuals generating and discharging activities and collective residuals handling and modification activities are induced to adopt physical measures for reducing discharges.

All implementation incentives require some institutional arrangement which provides authority to one or more agencies to impose the incentives on the activities. Because of the cross-cultural context of and the limited resources for the study, explicit analysis of implementation incentives was not included. However, suggestive examples of implementation incentives related to the specific physical measures previously discussed were identified and are described below.

Fuel substitution to reduce SO_2 and particulate discharges could presumably be implemented by communal and/or republic administrative regulation, i.e., a requirement that particular activities use fuels with maximum limits on sulphur and ash contents. A corollary incentive would be for the republic government to provide desulphurized fuel, perhaps with some degree of price subsidy. Similarly, the installation of cyclonic filters could be required, by regulation at the communal and/or republic level, on all boilers exceeding a specific capacity and using solid fuel. Alternatively, an effluent charge could be levied by the republic or the communes on each kilogram of sulphur and each kilogram of particulates discharged by each activity. The charge would then be a cost of production, similar to the costs of other factor inputs, such as raw materials and labor. An enterprise would seek to minimize total production costs--by changing fuel, installing cyclonic filters, changing production process. If the charge were sufficiently high, it would induce the adoption of one or more of the physical measures feasible for the particular activity. In addition the communal or republic governments could provide loans to, or make credit arrangement for, enterprises to help finance the installation of cyclonic filters or other RDR measures to reduce SO_2 and particulate discharges. Whatever physical measure is adopted requires both a monitoring and inspection system to check performance and apply sanctions for non-performance.

With respect to reducing discharges of HC and CO from vehicular movements, implementation of carburetor idling adjustment would require a system for monitoring and maintaining the performance of all gasoline-fueled vehicles in the five-commune area. Such monitoring and maintenance could be combined with existing regular safety monitoring and maintenance in communal inspection stations. Implementation of the Park and Ride measure to reduce discharges of HC, CO, and NO_x requires inducing individuals to forego the use of individual vehicles in the central city. This could be accomplished by adoption of some combination, or all, of the following incentives: (1) decrease the number of available parking spaces in the central city; (2) install meters on all parking spaces in the central city; (3) increase the metered parking rates; and (4) increase the frequency of monitoring the meters, accompanied by enforcement of fines for illegal parking.

Construction and operation of on-site wastewater treatment plants to reduce TSS and BOD_5 discharges from the twelve industrial plants, which are not feasible to connect to the municipal sewer system, could be accomplished by a regulation at the commune level requiring the construction of such plants by each enterprise, accompanied by a system of random inspections by a communal agency to check that the plants were being properly operated and a system of fines for failure to meet the discharge standards. Alternatively, as for SO_2 and particulates, an effluent charge could be levied on the discharge of each kilogram of TSS and BOD_5. If the charges were sufficiently high, the enterprises would be induced to adopt some physical measures for reducing their discharges. Communal and/or republic governments could provide loans to or make credit arrangements for the enterprises to help finance the installation of the physical measures. With respect to the regional wastewater treatment plant, the communes could proceed to implement its construction and operation in the same manner as with other infrastructure facilities.

Institutional Arrangements

As stated earlier, REQM involves a set of activities which must be performed in order to produce and maintain the desired product of improved AEQ. In any society those activities are likely to be performed by a set of institutions, agencies, enterprises, operating at all levels: local, metropolitan, regional, national. No single agency will perform all activities. The particular institutional arrangement for REQM will inevitably and necessarily reflect the sociopolitical, economic, and cultural systems of the society. It was impossible, in the Ljubljana area study, to undertake an extensive analysis of the institutional/governmental structure of loci of authority for imposing implementation incentives. However, some of the factors relevant to institutional arrangements for REQM in the Ljubljana area are suggested below.

First, at present (1976), no agency specifically responsible for REQM exists in Yugoslavia, although a number of communal, republic, and national enterprises/agencies carry out component parts of the REQM task. Second, REQM--and the ultimate responsibility for achieving and maintaining environmental quality--is delegated to the communes. The primary (virtually sole) role at present of the republic and federal governments is the establishment of "guideline" laws. Third, communal workers' management organizations[16] must initiate, develop, and enforce their own REQM strategies under federal and republic "guideline" laws. Fourth, the communes, as previously cited in several examples, have the potential to aggregate in a workers' management framework to form regional management bodies whose jurisdictional boundaries could coincide with relevant REQM problems.

The critical link in implementation is to induce communes to action, both to impose implementation incentives on individual activities and to undertake various collective measures for residuals handling and disposal. Because the communes must provide their own resources to affect AEQ, there may be little incentive to action unless serious AEQ conditions are perceived, as may be the case where many receptors in the five-commune area perceive significant damages to human health and property from adverse air quality. There may be little incentive to action because the costs to improve AEQ may be perceived as being too high and/or the resources available to deal with an REQM problem may be insufficient or non-existent at the communal level. Given the decentralized structure of the Yugoslav society, there may be no funds and/or no mechanisms to channel funds from higher levels of government to the communal level, as is frequently done in more centralized systems. At present, neither the federal government nor the republic government has significant "carrots or sticks" to induce communes to act. Clearly, unless it is *perceived* in a society that there is an AEQ problem, there is no stimulus for institutional development.

Concluding Observations

The results of the study described herein have immediate utility for REQM decision making in the Ljubljana area, with respect to both individual elements of the analysis and the overall analysis. First, the information developed clarifies and resolves several issues concerning: (a) the effect on ambient air quality of off-site central heating; (b) the relative contributions of industry and residences to air pollution as measured by ambient SO_2 and TSP concentrations; (c) the effects of connecting all activities in the five-commune area to sewage treatment facilities during the next two decades; (d) the effects of reducing vehicular movements on discharges of HC, CO, and NO_x in the five-commune area and the costs of such reductions.

Off-site central heating is not the key to reducing ambient SO_2 and TSP concentrations, as was widely hypothesized. It would in fact increase SO_2

discharges in the region and only slightly decrease particulate discharges, at very substantial costs. Contrary to the widespread opinion that residences have the major effects on SO_2 and TSP concentrations, industrial activities generally appear to have the more significant effects. With respect to reductions of TSS and BOD_5 discharges, significant differences in terms of both marginal reductions in discharges and related RDR costs were found to be achieved by the sequential addition to the municipal sewer treatment plant of feasible connection areas in the five communes. With respect to reductions in discharges of HC and CO, the two physical measures applied to vehicular movements achieve only small decreases in such discharges relative to what had been anticipated. This suggests: (1) that major changes in the transportation system within the five-commune area will be required if significant reductions in HC and CO discharges are to be achieved; and (2) that RDR measures directed at energy conversion in other activities will be necessary if a significant reduction in NO_x discharges is to be achieved, because NO_x discharges from vehicular movements account for less than 10 percent of NO_x discharges in the area.

Second, the study provides a data base and a methodology for continuing and improving REQM analysis for several of the agencies which became involved in the course of the study. For example, the municipal solid residuals collection agency intends to use data generated in the study to improve routing schedules for collecting mixed solid residuals and to investigate feasible incineration technologies for possible future adoption.

Third, even though a limited number of physical measures to reduce discharges of residuals were analyzed, especially with respect to industrial activities, the results show what degrees of reduction in the discharges of four major residuals--SO_2, particulates, TSS, BOD_5--can be achieved for different total expenditures. Such information, in addition to the perceived relative importance of those residuals, provides a basis for selecting an REQM strategy.

Fourth, a common view in the Ljubljana area is that ambient environmental quality can only be improved and maintained at the expense of economic development. Depending on the physical measures applied, this often is not the case, as detailed analyses of residuals management in industry have shown.[17] This point is made because of the view expressed in meetings with the Slovene Republic Committee on Air Quality, and in meetings with other groups, that reducing the discharge of residuals from industrial operations can only result in decreased productivity at increased costs. In fact, having to "rethink" the combination of factor inputs to produce a product or service--as a result of having to reduce residuals discharges--has led in many cases to decreased production costs, i.e., more efficient overall production.

Because this is an important issue, the annual costs for each physical measure applied to each relevant industrial category were calculated. Then, for each industrial category, what percentage these annual costs were of the wholesale market value of product outputs for the category in the base year, 1972, was then computed. The largest percentage found was 1.3 percent for substituting light liquid fuel for coal in the pulp and paper industry. All other percentages for industrial categories were less than 1 percent, indicating that industrial activities should be capable of paying the REQM costs with little impact on their competitive positions. This is particularly true because the costs for industrial activities were based solely on end-of-pipe and fuel substitution measures. Very often there are in-plant changes which achieve the same reduction in residuals discharges with substantially lower costs.

In summary, three basic analytical steps are required to enhance the utility for decision making of the work already accomplished. First, specific implementation incentives and the necessary institutional arrangement(s) to impose them must be identified for each of the physical measure-activity category combinations delineated in the study. This can only be done by the Yugoslavs who have the understanding of the political structure of Yugoslav society. Second, each physical measure/implementation incentive/institutional arrangement should be evaluated on the basis of the criteria delineated in Chapter I and the relative weights attached to those criteria by Yugoslav authorities. This would provide a rigorous basis for the selection of REQM strategies. Third, the analysis should be extended to consider changes in the area's economy which are occurring and will continue to occur. In the analysis performed in the study, it was necessary to assume static conditions for a base year (1972). The results of the analysis using this information may be acceptable for decision making perhaps through the mid- or late- 70's. After that time the changes in a rapidly expanding economy would require reanalysis, because of the probable changes in the spatial patterns of activities, in technology, and costs.

The present study of REQM in the Ljubljana area has demonstrated the utility of the REQM framework and of the systematic analysis of REQM problems, as well as having provided specific data for use in decision making. But more important, the study should be considered as the beginning of an ongoing, continuous activity of REQM analysis, which would continually provide information on which to base decisions for improving AEQ in the Ljubljana area consistent with the goal of continuing economic development.

Footnotes and References

[1] See Walter O. Spofford, Jr., Clifford S. Russell, and Robert A. Kelly, Environmental Quality Management: An Application to the Lower Delaware Valley, RFF Research Paper RP-1 (Washington: Resources for the Future, 1976).

[2] The full report of the Ljubljana REQM case study is, D.J. Basta, J.L. Lounsbury, and B.T. Bower, Analysis for Residuals-Environmental Quality Management: A Case Study of the Ljubljana Area of Yugoslavia (Washington: Resources for the Future, 1977).

[3] A commune is similar in size and function to a county in the U.S. See Zelena Kniga o Ogrozenosti Okolja v Slovenijil (The Green Book of Environmental Pollution in Slovenia) (Ljubljana: Zavod za Statistiko, SR Slovenije, 1971).

[4] Statistical Letopis SR Slovenije (Statistical Yearbook for Slovenia) (Ljubljana: Zavod za Statistiko, SR Slovenije, 1974).

[5] Ljubljanska Banka is the largest bank in Slovenia and presently finances more than 70 percent of all investments in Slovenia.

[6] The analysis for estimating liquid residuals generation coefficients showed that although wastewater discharged per inhabitant and the measured concentrations of BOD_5 (g/m^3) and TSS (g/m^3) in discharge streams differ between single-family and multi-family residences, the total BOD_5 and TSS discharges per inhabitant per day were the same. No attempt was made to evaluate infiltration loading due to garden and yard watering in single-family residences.

[7] Coefficients developed in the United States for steel industries by process type may, in fact, not be applicable to Yugoslav steel mills for the same process types because of different product output mixes and raw material inputs.

[8] Ventilation refers to inflows of air masses to, and outflows of air masses from, the basin.

[9] "Statjob" statistical package developed at the University of Wisconsin.

[10] See Chapter VII in Basta, D.J. et al, op. cit.

[11] For a description of water quantity/quality in the Ljubljana area and the Upper Sava Basin, see Chapter VII in Basta, D.J., et al, op. cit.

[12] See Polytechna-Hydroprojekt-Carlo Lotti and Co., The United Nations Study for the Regulation and Management of the Sava River in Yugoslavia, B/11-Water Utilization Plan, United Nations Publications, Prague-Roma, April 1972.

[13]It should be stressed that the analysis of costs discussed herein relates to economic analysis, i.e., the societal investment decision, not to financial analysis, i.e., the particular set of borrowing procedures and repayment rates relating to the means by which each of the activity categories would finance the specific physical measures.

[14]Ljubljanska Banka, March 1975.

[15]Only low energy content lignite and brown coal are in good supply. Liquid fuels, most of which are imported, are not in supply ample to satisfy greatly increased use.

[16]In the socialist economy of Yugoslavia, each enterprise in the social sector is operated in a workers' management framework which employs some form of self-government and profit sharing.

[17]See B.T. Bower, "Studies of Residuals Management in Industry", in E.S. Mills, ed., Economic Analysis of Environmental Problems (New York: National Bureau of Economic Research, 1975) pp. 275-320.

Chapter 6

CASE STUDY OF USING SYSTEMS ENGINEERING METHODS IN THE ENVIRONMENTAL POLLUTION RESEARCH PILOT AREA OF OSTRAVA, CZECHOSLOVAKIA

B. Řezníček

The purposes of this chapter are: (1) to give general information on the pilot area, Ostrava, and its problems; (2) to report on the environmental pollution research project carried out jointly by WHO and the Czechoslovak Government, with particular attention being paid to the question of using modelling and simulation methods; (3) to describe the work done up to the present, indicating the complex problems requiring solution during the next stage of the research; and (4) to present ideas concerning future work in developing mathematical models and making full use of them in environmental quality management.

Introduction

At the end of 1974, the joint WHO/Czechoslovakian Project, which includes the research being carried out in the Ostrava Pilot Area, was halfway towards its goal. This paper should therefore be considered as a case study of a research program which is in the process of realization. Nevertheless it should be useful to discuss the problem and share experiences at this stage of the work, in order to increase our knowledge and contribute to the solution of the problem by clarifying a number of its aspects.

In the framework of its general nationwide research program, the Czechoslovak Government carries out a number of important research projects in the field of environmental quality management. Because WHO sponsors analogous environmental quality research programs, it was formally agreed between the Czechoslovak Government and WHO to take joint action beginning in 1972 and continuing for five years. Subsequently the results would be put into practice.

Research tasks in the so-called "pilot areas" belong in the category of explicitly applied research with net output. How to make use of the results of monothematic research in the selected pilot areas is studied when taking practical measures to achieve the specific goals of environmental quality management in the given areas. Within the framework of the joint project eight pilot areas on ČSSR territory (four in Bohemia and Moravia, four in Slovakia) are dealt with, the pilot area of Ostrava being one of them.

Characteristics of Ostrava Pilot Area

Delimitation of the Area

The pilot area includes the town of Ostrava, its environs, and some parts of two neighboring districts, i.e., an area around the town within a radius of about 20 kilometers. That is the delimitation of the pilot area from the administrative point of view. It is obvious that every problem dealt with in accordance with the joint project will have its own particular area of action. For example, for water quality management the whole basin of the river Odra which flows through the pilot area will be assessed. When studying the negative effects of gaseous residuals discharges from industrial sources, the adverse effects of noxious discharges on the neighboring forest will be taken into consideration.

Natural Features

Ostrava and its outskirts rank with the largest and most important industrial areas of the ČSSR. The Ostrava area is situated in the region where the river Odra flows from its upper reaches and spreads to the lowlands of Poland. In the southeast it is bordered by the chain of Bezkydy mountains, in the northwest by the mountainous massif of Jeseniky. This geographical position of Ostrava causes the air to flow from the southwest to the northeast. The proximity of the Bezkydy and Jeseniky mountainous massifs affects the general direction of the prevailing winds as well as the local air circulation, causing aggravated aeration, many dead-calm periods, and development of fogs and of inversion conditions. In general, the basin of the Ostrava region shows a very insignificant vertical zoning. This fact influences in a positive way the possibility of developing mathematical models of the atmosphere using a homogeneous wind rose for the whole area.

From the hydrological point of view Ostrava is situated in the upper basin of the river Odra, which forms a natural backbone of water in the region. Unfortunately, only a small part of the basin is included in the pilot area, where the resulting lack of water resources represents the main limiting factor in studying water management there. The Odra river basin, with its sloped terrain, offers very favorable conditions for erosion and rapid flow of water with all the negative consequences involved, such as instability of the seasonal rate of flow, floods, and water shortages.

The disadvantages of the natural conditions from the point of view of hydrogeography have been for the most part improved by the construction of dams and reservoirs to provide water for drinking and for industrial purposes, such as Těrlice Dam. In spite of all the negative factors mentioned above the Ostrava region has one big advantage: it is situated in the upper basin. Therefore it has constant access to the source of high-quality water, on the condition that it is able to compensate for the seasonal flow variations. The interdependence of air quality and water quality management appears to be very important, in particular in the case of possible damages suffered by the Bezkydy forests due to the noxious discharges from local power plants, although their stacks comply with the hygienic standards which have been established.

Environmental Quality Management in the Ostrava Region

Background. As in other industrial centers, the environment has been degraded as a result of various activities. Air quality is adversely affected by sulphur dioxide, coal and ore dust, fly ash and exhaust gases; lowered water quality affects recreational activities; and the whole impact of the conditions of life in a modern industrial town on the health of the population is deleterious. In the last twenty years industrial development in the Ostrava region has reached a very high level. This extensive industrialization has had a number of negative effects, such as a continuous increase in the fallout of dust, increased discharge of industrial wastes to water courses, deterioration of landscape because of mining activities, and considerable growth in solid residuals generation, both industrial and municipal, and disposal on dumps.

When compared with other industrial regions of ČSSR, Ostrava has already made an important step forward. In the early sixties the administration dealt with the environmental quality problems of the Ostrava region, whereas in the other regions (including Prague, the capital city of the ČSSR), the necessity for adequate measures to be taken was not realized until the seventies. Some of the most important regulations concern: (1) the elaboration of an inventory of the sources of residuals discharges; (2) the making of environmental protection obligatory as a priority task for all decision makers; (3) cooperation of industry with local authorities (national committees) in environmental quality problems on a contractual basis; and (4) requiring those responsible for the most important sources of adverse environmental quality impacts to undertake specific studies on how to improve the situation stage by stage and to put the proposed measures into practice. As a result of these regulations an unprecedented result was achieved: by 1972 in the town of Ostrava the dust fallout had decreased by 70 percent. In that respect Ostrava has set an example for any future activity in the field of environmental quality management, and it may be said that the environmental quality issue in the Ostrava region meets general understanding and support. The good results noted above were achieved thanks to the adequate and systematic effort of the local authorities, that is, the Regional National Committee.

The substantial reduction in dust fallout was achieved by using traditional methods of management and accounts without taking advantage of new instruments of systems analysis and synthesis. However, the regional authorities are aware of the necessity for intensive development in the field of environmental quality management and have therefore established close contact with the UNDP/WHO Project. The ten-year effort to better the human environment has created in the Ostrava region a favorable atmosphere for using new and more sophisticated approaches to a number of problems, such as elaboration of materials and energy balances--including residuals generation and discharge--using modelling and simulation.

Economic Importance of the Ostrava Region

It is obvious that Ostrava in its future development will continue to maintain its character as an urban center with a high concentration of industrial production. The town of Ostrava itself is typical of one that has grown with the development of industry, as is shown in table 6-1.

Table 6-1. Number of Inhabitants of the Town of Ostrava

Year	1848	1880	1900	1946	1973
Number of inhabitants	1963	13400	30100	176600	292000

Environmental Quality of Ostrava at the Beginning of the Joint Project

In view of the scope and purpose of the Rotterdam meeting, this paper will deal first with the atmosphere as one of the subsystems of the environment. This has been given a high degree of priority in the pilot area of Ostrava. The problems of water quality management, of solid residuals disposal, and recultivation of lands that have deteriorated from the activity of man, modelling of which is still at the initial stage, will be discussed later.

As already mentioned, Ostrava managed to stop the decrease in air quality from particulate discharges. From 1962 a continuous decrease of the respective values can be observed, as shown in figure 6-1. The curves demonstrate: (1) the continuous decreasing trend of measured values; (2) an occasional increase of the respective values due to the construction and putting into operation of a new plant before the necessary measures

Figure 6-1. Relative Values of Particulate Concentrations in Ostrava, 1960-1972

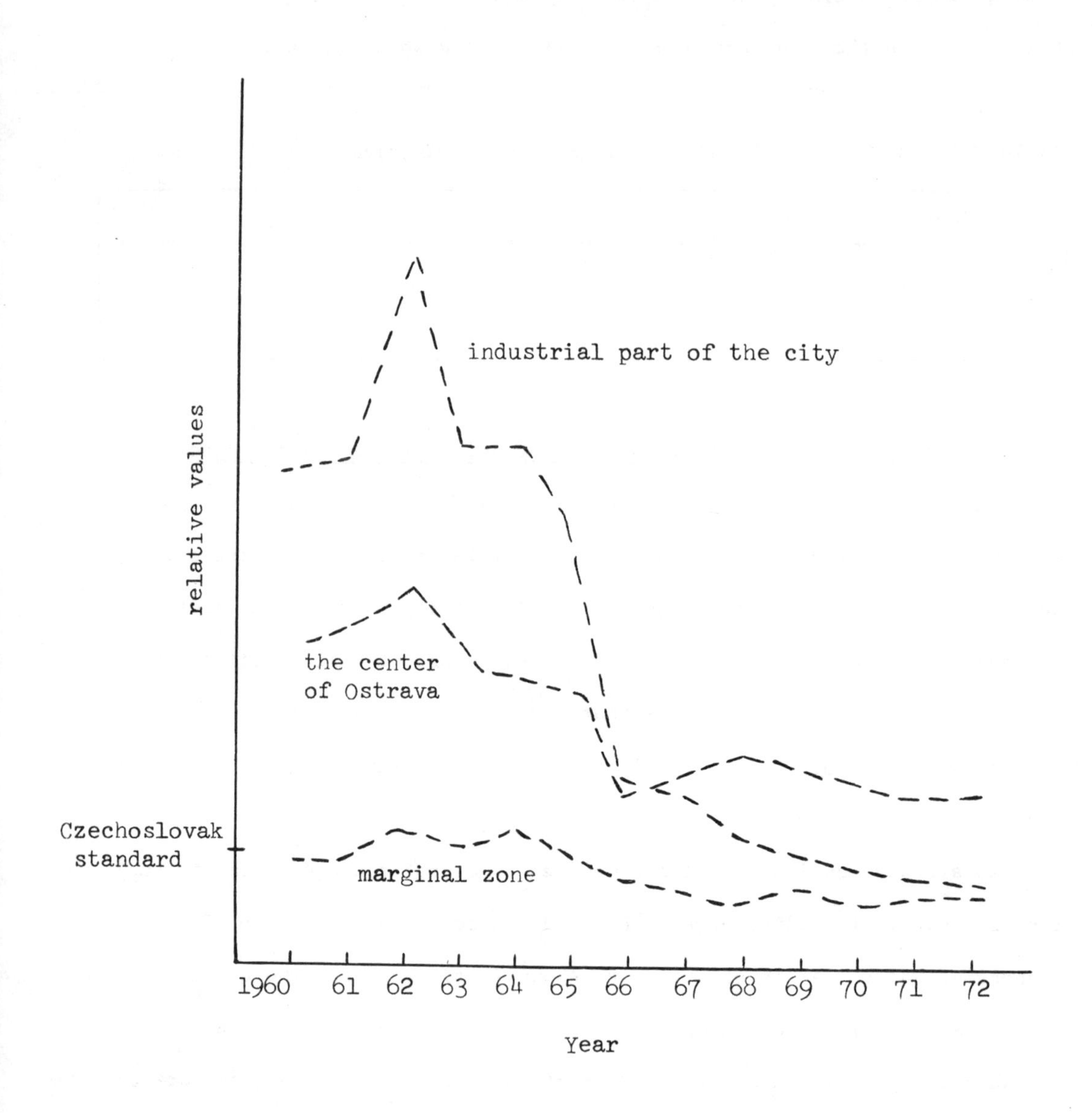

for reducing particulate discharges were taken; (3) the fact that not all industrial enterprises comply with the Czechoslovak hygienic standard; and (4) the necessity to continue the fight against air quality deterioration using new and more efficient instruments and methods (mathematical models).

Less favorable is the situation concerning the problem of sulphur dioxide discharges. Today's worldwide energy crisis has seriously aggravated the energy balance of most countries all over the world. Therefore it is unlikely that recommendations for the use of "cleaner" fuels instead of those with a high sulphur content--recommendations resulting, for instance, from the simulation of dispersion of SO_2 in the atmosphere--could be put into effect. However, the results of simulations using dispersion models can influence decisions as to where a projected plant should be built, and what characteristics of desulphurizing equipment or stacks should be.

Before giving its approval to a proposed study, the Project Institute in Prague calculates the expected air quality with the help of the dispersion model. This kind of model, though uncomplicated, is not at the permanent disposal of engineers and technicians of the local administration.

As a result of growing industrialization a decrease in ambient water quality has been observed. This has been due to the fact that during the first stage of industrialization after 1945 treatment of wastewaters was carried out very superficially. However, between 1960 and 1970 an extensive research project and investment program sponsored by the Government was launched for the construction of municipal and industrial wastewater treatment plants. This program was very successful. That is the reason why the problem of water quality management has not been given high priority in the Ostrava region up to the present.

Experts dealing with the problem of the pilot area recently decided that rather than wait to model water quality in the water course until several new plants in the Odra river basin had been completed and the character of liquid residuals discharges and the resulting water quality had more or less stabilized, the work should start immediately. If the model of the water course were not developed at the present stage, it would not be possible to establish, with the help of simulation, the expected ambient water quality in the future when the projected industrial plants would be in operation, and the possibility of influencing the decision-making process would be lost.

Another environmental quality problem is the deterioration of cultivated lands because of subsidence from the deep extraction of coal. This problem will be linked with the question of solid residuals disposal and dealt with as one research task. The work was expected to start in 1976 in addition to that of the UNDP/WHO Project.

Goals, Method, and Organization of the Research

Goals and Method

As mentioned before, the authorities of the Ostrava region give full support to the research work and they expect to make full use of the results. In close cooperation with them, the goals, the priorities, and the methods were established. The following were the agreed upon priorities: (1) development of a system for air quality management; (2) experimental solution of problems of soil recultivation and solid residuals disposal; and (3) research into management of water courses with respect to water quality.

As the working method of systems engineering was chosen, especially in cases where it was necessary to identify and define the interrelationships among the particular subsystems of the environment which are still in most cases considered as autonomous components, it was decided to proceed from the very beginning to analyze and synthesize mathematical models and their simulations.

Organization of the Research

The research network on the Ostrava pilot area dealing with environmental quality management is fairly well organized and developed, but it is more or less oriented towards traditional technology. Therefore, in order to create the necessary conditions insuring that the research would be carried out successfully, a working group of representatives of all bodies involved was established to participate in the common research tasks and to work as a team. The variety of levels and bodies represented (ministries, industrial and other enterprises, organizations and institutions) created obstacles that had to be overcome. In this connection the authority of the nationwide research program arising from the so-called "Program P 16" sponsored by the Ministry of Technical Development and Investments, as well as the authority of the UNDP/WHO Project, were of great importance and help. Another important factor was the feeling of regional solidarity of the particular research institutes, and their sincere efforts to contribute to the solution of the environmental quality problems of their regional capital.

In 1973 all institutions dealing with the problems of air quality management (using mathematical modelling methods) were linked. This involved:

the hydrometeorological institutions; institutes of systems engineering; institutions of air hygiene; and air quality inspections body. In 1974 it was intended to organize close cooperation among specialists of other institutions dealing with modelling and simulation of water quality--technicians, systems engineers, hydrological experts, inspectors of water quality protection, etc. The following year the institutions dealing with soil recultivation and solid residuals disposal will be linked. Even in this case it is intended to make use of a mathematical model of residuals generation (input-output model) as an element helping to consolidate the work of the team.

Organization and Management of Interdisciplinary Research with the Use of Mathematical Models

The head of an interdisciplinary team dealing with environmental quality problems is in a very difficult position. At present there are few instruments which could help him in his management activities. Experience acquired in managing a working team dealing with the modelling of continuously operated power and chemical plants (in the universal problem-oriented simulation language PACER) lead us to the conclusion that even the simplest mathematical model represents a very important basis for the work of any interdisciplinary team. A mathematical model helps toward: (1) a better understanding of the problems; (2) a better division of work among individual experts; and (3) an improvement in communication among team members. The last is represented in figure 6-2.

In the course of activities in the Ostrava pilot area the conclusion was reached that no one of the future environmental quality research tasks dealt with by an interdisciplinary team should be included in the working

Figure 6-2. Conceptual Representation of the Provision of Effective Exchange of Information in Interdisciplinary Research by the Common Language of Mathematical Models

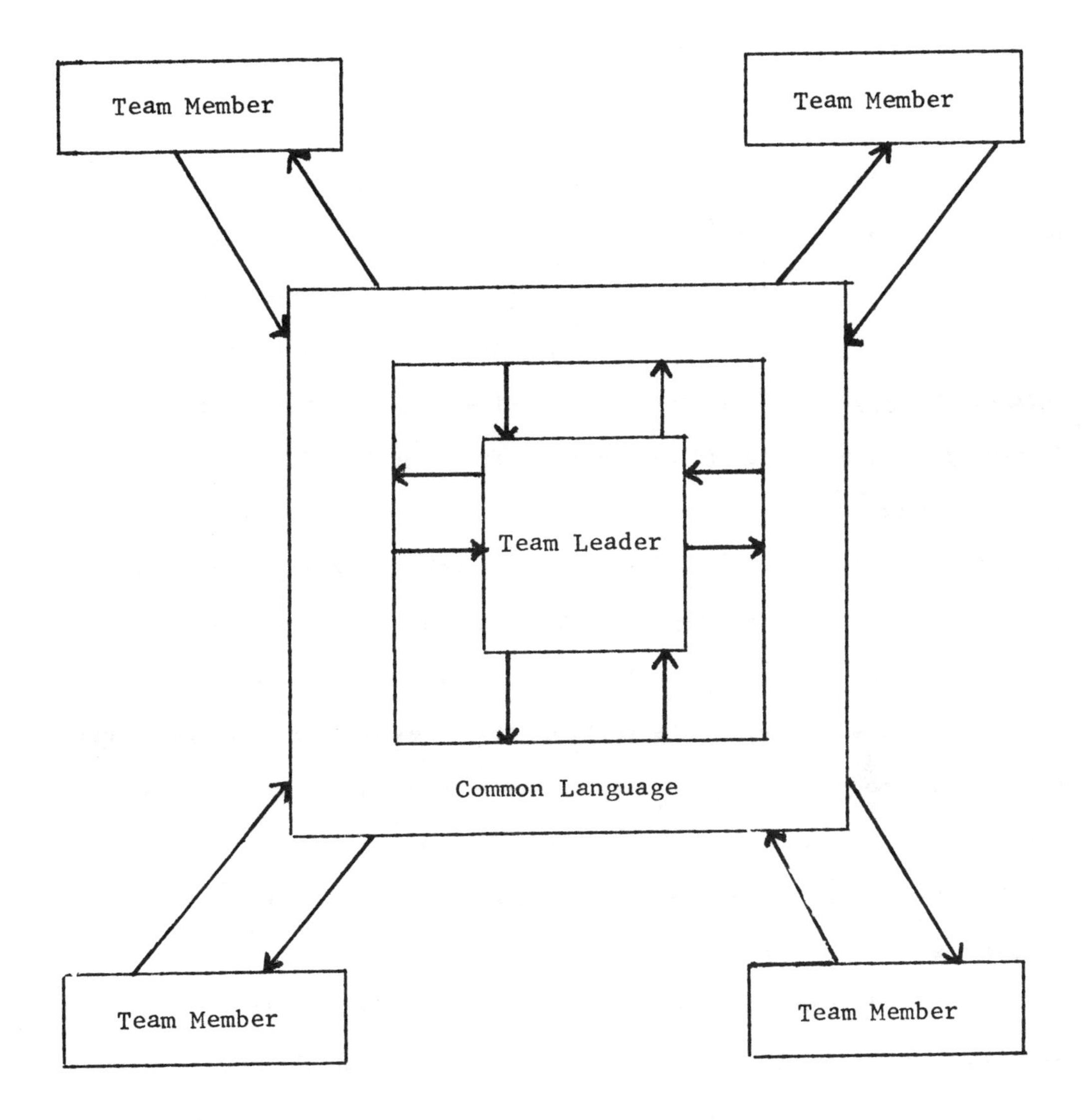

program unless it were based on a preparatory project of a mathematical model, even when not required by the objectives defined in the program of research activities. A mathematical model appears to be an important methodological aid which helps communicate the necessary information, promotes solution-oriented discussion, and helps the experts think over their work and orient it to an economic accomplishment of the tasks.

Progress to Date

At the first stage of the research work the team dealt with methodological questions and with drawing up an inventory of available information. This was necessary to acquaint the research workers with the problems and with their tasks within the team. A serious obstacle to the work was the appreciable shortage of technical and scientific information. To remedy this, valuable aid was received from the UNDP/WHO Project which, at the next stage of activities, placed at the team's disposal information and recommendations of UN experts visiting Czechoslovakia. In that way difficulties arising from the gap in the fundamental research was overcome.

In 1973-1974 attention was turned from the environment as a whole to its individual subsystems. The goals of the research work were accordingly defined with more accuracy. Since the beginning of 1974 one of the working groups of the team has dealt with the IPP[1] model. Colleagues in Bratislava have already modified the program and are ready to adapt it to our needs in the future.

The establishment of a working group in the Ostrava pilot area was motivated by the following objectives: (1) to activate the most appropriate of the modified versions of the IPP program with the help of a computer which is within easy reach of the group; (2) to provide the necessary input data; and (3) to develop the mathematical model of air quality in the pilot area and verify its adequacy. Later on the verified model could be used for elaborating a program for air quality management having in mind future industrial development and output.

The 1974-1975 program for the Ostrava pilot area aimed: to activate the dispersion part of the IPP model with the most appropriate modifications (in cooperation with the Czechoslovak Centre in Bratislava); to elaborate a study concerning the utility of water course modelling with some convenient ways of water quality modelling; and to prepare a study of modelling solid residuals generation and disposal.

The execution of these tasks will create the favorable preconditions for our future work in the application of mathematical models. Those models are expected to help: to elaborate and to transmit to the executing institution the dispersion model of the atmosphere as an aid to decision making; to define the problem of water quality management, its objectives, and to propose the basic principles of such activity in cooperation with the project executing institution; and to insure that the research in the Ostrava pilot area will be continued within the framework of the nationwide program of science and research development after the UNDP/WHO Project is finished.

When dealing with these new tasks which are focused on the stabilization and improvement of the components of environmental quality in the Ostrava region, and proposing a system of environmental quality management, use will be made of mathematical models. Great benefit will be derived from previous experience in the use of input-output models in environmental quality management. (This matter was dealt with in 1970-1971 in addition to the work of the UNDP/WHO Project.)

Models of residuals generation have not been used until now because it was preferred to concentrate efforts on the dispersion models, to prove their utility and efficiency, and only then to turn attention to the former models. Although aware of the growing demand for data collection and evaluation it is felt that a project with an appropriately structured data basis must hold precedence.

Summary and Conclusions

In this paper the results of the first stage of work in the Ostrava pilot area and the problems faced have been discussed. Above all it is a question of the number of highly specialized research workers being limited. But there also arises the question of the relatively high degree of inertia in man's thinking and his reluctance to accept new knowledge and adapt it to all scientific disciplines.

In Czechoslovakia the problem of using the "systems" way of thinking and systems approach to environmental quality issues had been dealt with before the UNDP/WHO Project had been launched. Linking some of the tasks of the Czechoslovak research program in the Ostrava pilot area with the UNDP/WHO Project helped to define the problems better, to present them more clearly, and to give initial consideration to the possible solutions. It became evident that making use of the modelling approach offered some additional advantages in the management of the activities of an interdisciplinary team.

This paper has dealt mainly with the different approaches to the research. It is hoped that on some other occasion there will be the possibility to inform a meeting similar to this Rotterdam meeting of the specific results of the work described herein, which work is to be partially finished in 1975 and completed at the beginning of 1977.

Footnote

[1]TRW, Inc., Air Quality Implementation Planning Program: Volume I--Operators Manual, Volume II--Programmers' Manual, U.S. Environmental Protection Agency, Washington, C.C., November 1970 (National Technical Information Service, Springfield, Va., 22151, accession nos. PB 198-299 and PB 198-300).

References

1. Reports on the Tasks of Environmental Research in the Ostrava Pilot Area, Ostrava, 1972-1974.

2. B. T. Bower, "Residuals-Environmental Quality Management in the Urban Environment," presented at the conference on Environmental Management of an Urban World System (Noordwijk, The Netherlands, September 1971.)

3. G. M. Jenkins and P. V. Youle, Systems Engineering and Its Unifying Influence in Industry and Society (London, 1971).

4. B. Řezníček, "Systems Engineering and Environmental Sciences," Symposium on Systems Engineering Application in the Tertiary Sphere (Mariebad, October 1973).

5. B. Reznicek, V. Reznickova, and R. Bogusovsky, "Rationalizing the Techniques for Rationalization," Symposium on Systems Engineering Application in Continuous-Flow Processes (Mariebad, October 1972).

6. C. S. Russell, W. O. Spofford, Jr., and E. T. Haefele, "Environmental Quality Management in Metropolitan Areas," presented at the conference on Urbanization and Environment (Copenhagen, June 1972).

Chapter 7

WHAT THE ROTTERDAM SESSION HATH WROUGHT

Blair T. Bower*

Introduction

The background paper by Kneese and Bower (Chapter 2) provided the starting point for the discussions at the Rotterdam meeting. This chapter attempts to capture the salient points made and questions raised in, and the conclusions stemming from, the discussions. These points, questions, conclusions are organized in three categories: (1) general points; (2) model formulation and mathematical technique; and (3) potential use and users. The three categories are not completely unambiguous; some of the points could be placed in more than one category. (In fact, some repetition exists.) Further, the points are not necessarily listed in order of importance.

To provide a common basis for the explication of the points, figure 7-1 depicts regional REQM modelling. A regional REQM model is comprised of a set of models relating to economic activities distributed over space in some region: residuals generation by those economic activities; costs of on-site methods of reducing by various degrees residuals discharges from each activity; costs of collective methods for modifying residuals and for modifying AEQ directly; environmental models which translate the spatial pattern of residuals discharges and direct environmental modification into the resulting spatial pattern of AEQ; and one or more sets of AEQ standards and/or damage functions, the latter translating the spatial pattern of AEQ into

* Particularly helpful comments were received from Clifford Russell on the original draft of this chapter.

Figure 7-1. Regional REQM Model

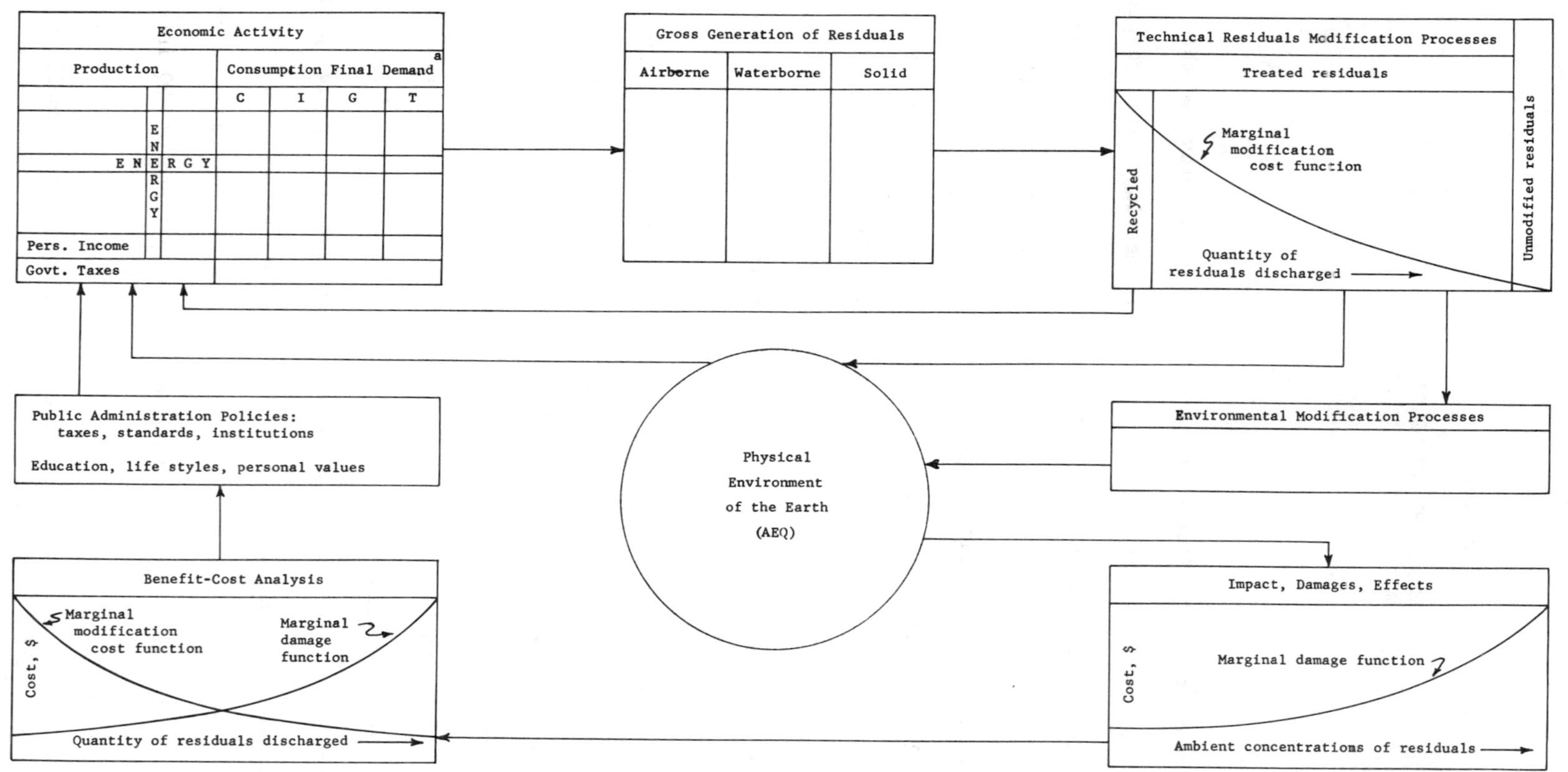

a C = consumption, I = investment, G = government, T = trade

Source: John H. Cumberland and Bruce N. Stram, "Empirical Results from Application of Input-Output Techniques to Environmental Problems," in Karen Polenske and Jiri Skolka, eds., Proceedings of the 1974 Vienna Input-Output Conference (Cambridge, Mass.: Ballinger Publishing Company, 1975).

damages to receptors--man, plants, animals, materials. The set of models is used for the analysis of REQM strategies.

General Points

1. Modelling REQM on a regional basis is useful, but careful consideration must be given to existing and possible future institutional structures and political jurisdictions. Where the boundaries of political jurisdictions do not coincide with residuals management boundaries, e.g., "airsheds," regional REQM modelling/analysis may have to consider explicitly imports of residuals into, and export of residuals from, the designated area.

2. Regional REQM modelling is not always necessary. Situations exist where AEQ is so bad and the sources of the adverse conditions are clearly known, that little in the way of a formal model is necessary to determine a local REQM strategy, at least for a politically realistic "first cut."

3. If the objective in a given context is to achieve an actual improvement in AEQ, the weakest link in the "sequence" of activities leading to such a result may well not be, and in fact often is not, the analytical capability, i.e., regional REQM modelling, but rather the capacity to select and impose implementation incentives to induce adoption of the physical measures to improve AEQ. (See #2 of the section on "Potential Use and Users.")

4. If the objective in a given context is to achieve an actual improvement in AEQ, the analysis must include: sources of residuals discharges; the various physical measures for reducing the discharge of those residuals; the implementation incentives for inducing the activities to reduce their discharges; and the institutional structure (mix of governmental agencies) with authorities to apply the incentives to the activities, including col-

lective activities, to induce the adoption of the "optimal" set of physical measures to achieve the desired AEQ. The linkages among these elements are illustrated in figure 7-2.

5. In regional REQM modelling to date (1975), only a few indicators of AEQ have been used, such as concentrations of sulphur dioxide, particulates, dissolved oxygen, fish biomass. For SO_2 and particulates the rationale for the target concentrations is protection of human health. But these indicators may well not reflect the most important residuals affecting human health, nor is it known what level of "protection" in relation to human health is really desired by any society. Because man as a "whole" is the target--in terms of total body burden--more consideration needs to be given to what specific indicators of AEQ should be included in regional REQM modelling. This is particularly true considering the number of new chemical compounds annually developed for use.

6. Regional REQM models are useful for:

a. helping choose goals with respect to REQM, i.e., level and distribution of AEQ to be sought; and

b. helping choose REQM strategies (physical measures + implementation incentives + institutional arrangements) for achieving a given goal. If a goal has already been clearly defined, application of regional REQM modelling is straightforward; if the goal has not been clearly defined, the application of regional REQM modelling itself becomes an input into defining the goals.

7. Regional REQM modelling must be designed in relation to the particular context or setting of the modelling, objectives of the analysis, modelling capabilities, availability of data, and availability of computa-

Figure 7-2. REQM Linkages

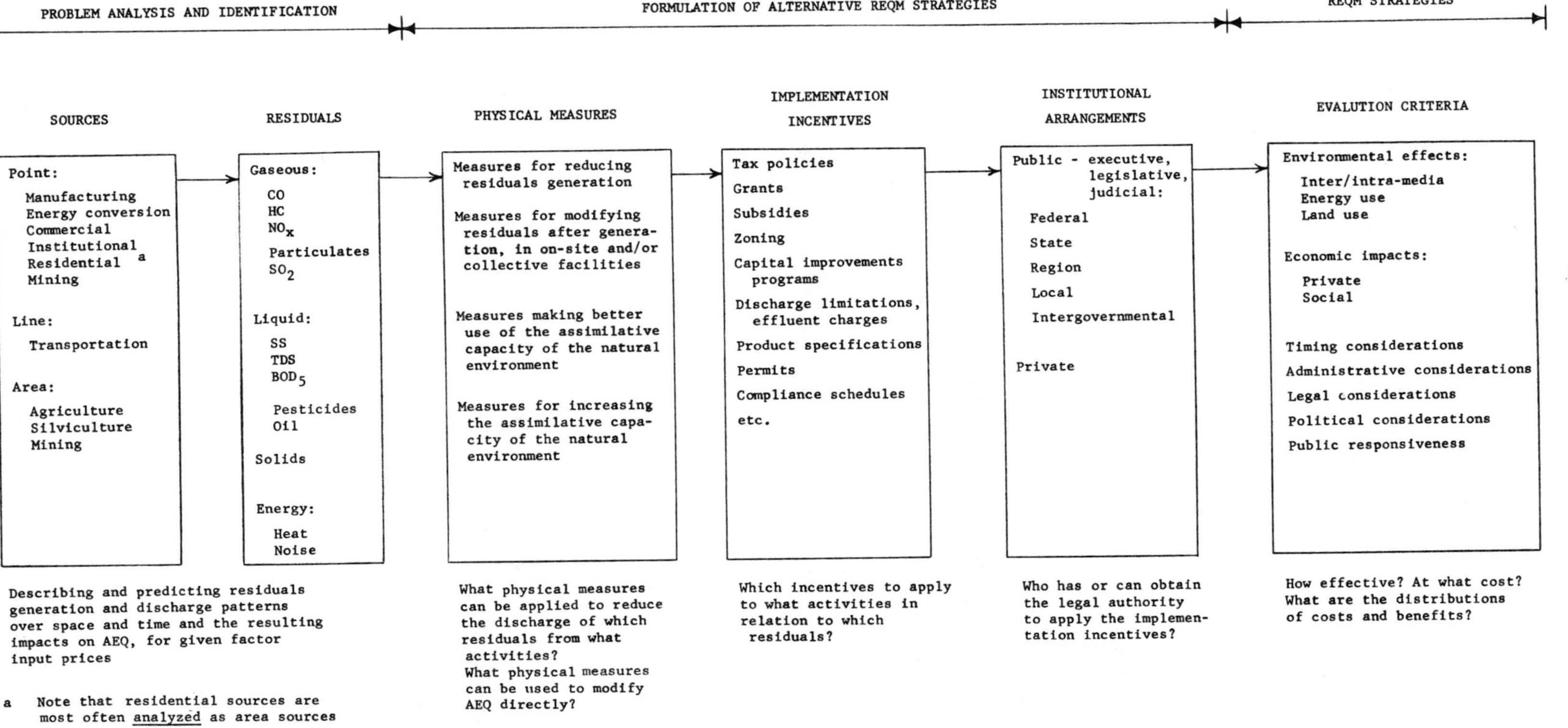

tional and analytical resources. Each application of regional REQM modelling requires the allocation of the available analytical resources to preclude over-refinement of any one element and to achieve the optimal overall output with the resources available, within the available time frame for the modelling.

8. For some residuals the globe is the relevant region. But no operational ways of analyzing such a region are available at present.

Model Formulation and Mathematical Techniques

1. Model complexity. The complexity of regional REQM modelling can be described in terms of three dimensions:

   a. scale (geographic scope);

   b. number of variables explicitly considered, i.e., decision variables; and

   c. time.

Figure 7-3 attempts to depict the extent to which regional REQM models have incorporated these dimensions, in contrast to optimal control models. The former have included more decision variables, greater geographic scope, but a more limited time dimension than the latter. Optimal control models have adequately included the time dimension, but not a large number of decision variables or any analysis of geographic scope.

The time dimension itself has two aspects: dynamic and stochastic, the former defined in terms of changes from year to year. The stochastic variations--diurnal, day-to-day, weekly, seasonal--are superimposed upon changing conditions from year to year, such as increasing or decreasing levels of economic activities, changing technology and factor prices, accumulation of residuals in the environment and the corresponding impacts

Figure 7-3. Complexity of Present Generation Models

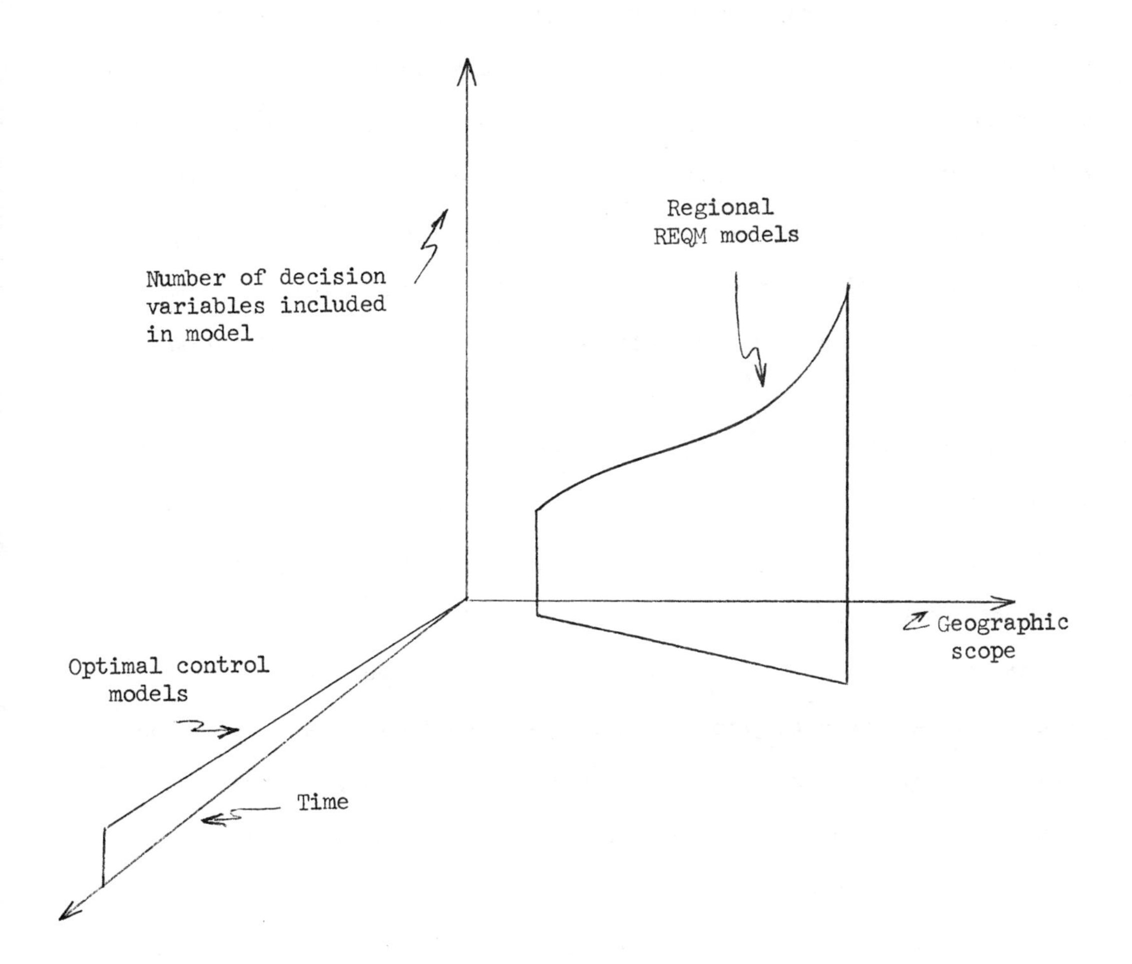

on ecosystems. What are the payoffs to, and problems in, increasing the complexity of regional REQM models? The latter question can be more readily answered than the former, because potential payoffs can only be dimly foreseen until such models have been constructed and the types of information they can generate determined.

(i) Present steady state type regional REQM models can be made more comprehensive by including activity analysis type models directly in regional and interregional input-output models. Procedures for doing so seem rather straightforward, so that achieving such expansion is primarily a matter of the resources devoted to the task. On the other hand, such expansion substantially increases data collection, data arrangement, and computational problems. It may well be more efficient to handle "trade" effects separately.

(ii) Including dynamic and stochastic aspects _integrally_ in large--and even not so large--regional REQM models is a much more difficult task. Until this is done, however, some important questions for REQM cannot be answered rigorously, for example, with respect to dynamic aspects, questions relating to chronic effects and accumulation of residuals in the environment. Stochastic aspects must be incorporated if the impacts of short-term residuals discharges are to be analyzed and REQM strategies to mitigate the impacts if necessary are to be developed.

(iii) The gap between optimal control models of intergenerational problems and the existing operational regional REQM models is very large. The former are more sophisticated and general but at present can handle only one or two decision variables. Nevertheless, there may be cases where such models can usefully be applied to residuals problems.

(iv) Conceptually, activity models can be incorporated into regional REQM models and regional input-output models which in turn can be linked to interregional industry models. Or national (macro) input-output models can be disaggregated into regions, where for each region a regional REQM model could be developed, including individual activity models, with subsequent feedback into the interregional and macro models. However, it is not clear to what extent the outputs from the mass of interrelationships involved in macro models are meaningful. If, among regions, differential capacities to absorb residuals are to be considered explicitly in location decisions, then some form of interregional model is essential.

2. Modelling cost. The cost of applying regional REQM modelling includes: the cost of data collection and arrangement; the cost of model formulation, calibration, and verification; cost of analysis with model; and the cost of making an incorrect decision if the model data are inaccurate and/or the model is inadequate. In some cases, such as for the Lower Delaware Valley, the cost of raw data was relatively low; in other cases, such as for the Ljubljana area of Yugoslavia, the cost of raw data was high because virtually no data existed. Tradeoffs between empirical data collection and model formulation can be made in relation to the type of decision involved and the relative importance, and the levels of uncertainty, associated with the different elements of the regional REQM model.

Preliminary sensitivity analysis can be of aid in determining how to allocate the resources available for regional REQM modelling.

3. Activity models.

a. Most, if not all, of the regional REQM models developed to date, including input-output models, are based on the levels and locations of economic activity and residuals generation and residuals discharge data which are 5-10 years old. How useful are such models in developing REQM strategies, given the fact that the world is continuously changing "underneath the analyst," i.e., changing technology, changing factor prices, changing product output specifications, changes in various nonresiduals-oriented governmental policies which indirectly affect residuals generation and discharge? What is the "lifetime" of a model?

b. Under normal conditions for all activities, there is very substantial variation in residuals generation and discharge, within a day, from day to day, weekly, seasonally. Generally, variations for gaseous residuals are larger than for liquid residuals. Further, there are additional variations resulting from start-up and shutdown operations and from accidental spills and breakdowns. In many cases the residuals generation and residuals discharge data used in regional REQM models are based on a few instantaneous samples. The variability in residuals generation and discharge raises questions about the adequacy of these data.

c. Existing regional REQM models generally have given inadequate consideration to nonpoint sources such as: urban storm runoff; washout from precipitation; dumping of used oil by individual car owners, garages, service stations directly into the environment--estimates of 20 percent to 40 percent of used oil being "disposed of" in this manner have been made for specific areas--and runoff from agricultural and silvicultural activities. In some regions such sources of residuals discharges are predominant. In such cases, little improvement in AEQ is possible unless the REQM strategies are explicitly directed toward these sources.

d. The efficiency of the REQM strategy developed by regional REQM modelling is a function of the range of options included for reducing the discharge of residuals into the various environmental media and for modifying AEQ directly. Dependence solely on "conventional waste treatment" measures for reducing discharges of residuals can result in costs 2 to 2.5 times the costs if various internal process modifications and materials recovery/by-product options are included, for a given raw material and a given product output. To the extent that changes in product specifications are possible, costs can be reduced still further. To what extent then should the following be included in developing regional REQM models: (1) existing range of technological options; (2) potential range of technological options, based on current research and development; (3) existing range of product output specifications; and (4) potential range of product output specifications? To what extent should alternative sets of final demands and "life styles" be considered?

e. Generalized cost functions for reduction in residuals discharges have often been used in regional REQM modelling. If discharge reduction costs are significantly site and plant specific, under what conditions are the generalized cost functions adequate? In developing cost functions there are also problems in attempting to consider explicitly scale economies and interactions among residuals, at least where linear activity models are utilized. The importance of neglecting these "real world" factors varies from case to case.

f. The locations of the individual activities in regional REQM models have generally been assumed to be fixed, except where two or more explicit alternative spatial patterns of economic activities have been separately analyzed. Allowing the site of an individual activity to vary, substantially complicates a regional REQM model. If site is to be a variable, this will likely come at the expense of less detail in the activity models and/or the environmental models.

g. Increasing concern has been exhibited with respect to many individual elements and compounds which have not been included in regional REQM models to date, i.e., heavy metals, organic chemicals, noble gases. The number and synergistic combinations of these are very large. Rather than attempting to include these residuals directly in regional REQM models and in interregional interindustry models, they might be analyzed in partial, side calculations using data from individual enterprise materials-energy balances.

h. The structure of activity models may also be conditioned by an objective of obtaining information on a specific factor input, such as energy. In such cases the models must be designed to identify the variable of concern explicitly.

4. Environmental models. The type of environmental model and the types of outputs from the model are a function of the objective of the regional REQM models. If the model is to be used for developing REQM strategies in a particular context, then the environmental model must fit into the economic-technologic management decision framework relevant to that context.

a. What environmental conditions are to be chosen for analysis in regional REQM modelling: long-term average conditions; seasonal conditions; short-term peak conditions (episodes)? Must the time frame of the meteorological conditions chosen be the same as the time frame for the hydrologic conditions? How can the analyses of several different sets of environmental conditions--for example, different seasons--be utilized?

b. For REQM, not only the mean annual environmental conditions must be assessed but also those which occur, for example, only one percent or a few percent of the time. This raises the problem of assessing the impacts of short-term loads on ecosystems, both where only a few components of the system are "wiped out" and where virtually all components are wiped out.

c. Global environmental models are relvant for particular problems, such as the SST, spread of DDT, oil on the oceans. As yet no operational models for such problems have been developed for use in REQM analysis. What proxies for such models might be used? How can the interactions between a regional model and "the world" be included in an analysis?

d. How to include the problems of accumulation and irreversibility into the environmental models of regional REQM modelling has not yet been "solved."

e. With respect to analyzing the problems of toxic residuals--where minimal knowledge exists with respect to their impact on and movement through the environment and various ecosystems and where thousands of new types of such materials enter the system each year--there is perhaps a need for development of some simple, qualitative environmental models.

f. Where an environmental model has been developed under conditions of essentially "no pollution," to what extent can it be assumed that such a model will actually predict the effects on AEQ of substantially increased residuals discharges and vice versa?

5. Intermedia relationships. A central issue in regional REQM modelling is the extent to which intermedia relationships must be analyzed simul-

taneously. Both individual industry studies and the Lower Delaware Valley regional study suggest that the linkages among the forms of residuals and the three media are real and significant, thereby requiring simultaneous analysis. However, in some cases the intermedia linkages may be small enough to be handled by ad hoc procedures, such as specifying particular contraints on one form of residual or specifying a particular method and level of modifying the residual.

6. Aggregation. The degree of aggregation/disaggregation in regional REQM modelling depends on: (a) the extent to which a regional REQM model is to be explicitly tied to a regional input-output model and/or to an interregional model; (b) the desire to reduce the complexity of the model to expedite analysis; (c) the desired level of accuracy of information in the components of the model, i.e., activity models and environmental models, where increased accuracy is representd by inclusion of more options for reducing residuals discharges in the activity models and increasing the number of grids or number of river reaches in the environmental models; (d) the method of solving the environmental models; and (e) the extent to which a particular activity-residual combination can be affected by an REQM strategy. (An example of the last is separating the generation of mixed solid residuals into specific types of used paper products and all other solid residuals where specific management options are available for the used paper products.) In regional REQM modelling there are possible trade-offs among all of these.

7. Simulation optimization vs. mathematical optimization. All models are simulations in the sense that some exogenous conditions are chosen, the model is run with the assumption that it mimics some part(s) of the world

and the behavior of the actors therein, and an "answer" comes forth. Mathematical optimization procedures provide some systematic method for finding a better-and-better answer when measured against the chosen objective. However: (a) there is never a guarantee that the "best" answer has been found if there are nonconvexities; and (b) the "systematic" method may be more or less efficient, and in the limit, may amount to nothing more than "brute force," so that the line between this and simulation optimization nearly disappears. The following considerations are relevant to the choice of analytical approach. There are trade-offs between the two approaches, in terms of the types of output information provided.

a. Mathematical optimization imposes a valuable discipline upon the regional REQM modeller.

b. If AEQ standards are the targets to be achieved by the REQM strategy, and there are a large number of discharges, a large number of options, and a large number of receptors, mathematical optimization will usually be the reasonable way to proceed.

c. If there are only a few major dischargers and a few options for reducing residuals discharges, simulation optimization may be easier and sufficiently efficient.

d. If there are many similar sources of residuals discharges and only one or two options for reducing discharges at each source, and the objective is defined in terms of required reductions or percent reduction of discharges by the sources--rather than in terms of AEQ--simulation optimization again may be easier and sufficiently efficient.

e. Large mathematical optimization models are expensive and difficult, unless linear.

f. Large, nonlinear models very often have multiple "optima." Or, the distributions of costs among the dischargers may be significantly different for approximately the same value of the objective function.

g. The linearity assumptions may or may not do great violence to the "real world."

h. Synergistic and antagonistic effects are more difficult to handle in mathematical optimization than in simulation optimization models.

i. Economies of scale or analogous increasing-return situations cannot be handled satisfactorily by any mathematical optimization techniques.

8. Sensitivity analysis. Given the essential impossibility of validating regional REQM models, it is important to know the sensitivity of the results (output values) to the various variables in the models, i.e., in the activity models, environmental models, cost models. Sensitivity analysis is also useful for the selection of REQM strategies, in that it can indicate which are the more important sources of residuals.

In some cases regional REQM models have been developed using present conditions, with the results predicted by the model being compared with observed (measured) values of AEQ. Where there are large differences between predicted and observed values, are the activity models, the environmental models, or both inaccurate? If the predicted values are spproximately equal to the observed, this still does not mean that the model will predict accurately for changed conditions, because the analyst might have been "lucky," in that errors in the activity models compensated for the errors in the environmental models.

9. Models of toxics. Regional REQM modelling has not yet been developed for toxics, carcinogens, mutagens, etc., because of lack of data on: generation and discharge; costs of reducing discharges; effects on ecosystems; and practical techniques for optimization analysis of large, dynamic sys-

tems. (To assess the impacts of such residuals requires dynamic models.) If it is assumed that large reductions in the discharges of such materials are necessary, whatever the costs, then simulation can be used in the analysis.

Potential Use and Users

1. A major problem in the application of regional REQM modelling is that of translating the outputs from the analysis into meaningful information for decision makers. This implies that knowledge exists on the part of the analyst with respect to the questions the decision makers wish answered. This may or may not be the case. For example, the decision makers may simply not be interested in AEQ problems; or they may be unable to formulate operational questions. Thus there is the broad and complex issue of what linkages can and/or should be developed between the analysts and the various users--governmental executive agencies, governmental legislative bodies, private interest groups.

2. Regional REQM modelling can be considered as part of the total REQM system, as depicted in figure 7-4. If the objective is in fact to achieve an improvement in AEQ, i.e., minimize the difference between observed AEQ and AEQ standards, then allocating further resources to what is already probably the strongest of the four links, i.e., regional REQM modelling, will not result in optimizing the <u>total</u> system. What is needed is to analyze the total system, to try to develop criteria for determining whether or not, for example, to add to the complexity of the regional REQM models if the information from the existing level of complexity is already not being used by the decision makers.

Figure 7-4. Representation of REQM as a System of Serially Connected Links

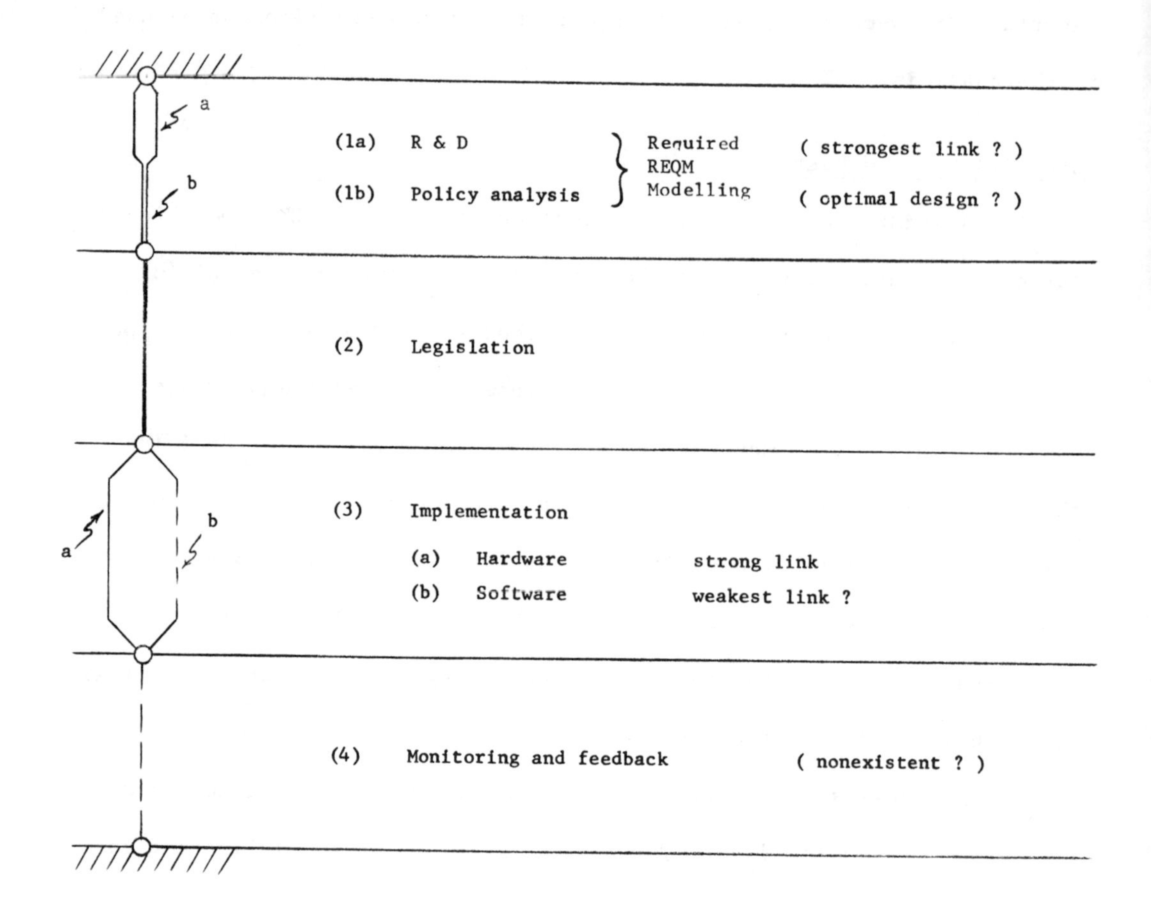

3. The objective of regional REQM modelling is to provide information to the decision-making process, not to substitute for it. For example, the research arm of the U.S. Environmental Protection Agency is using the regional REQM modelling approach to provide technical guidance to EPA program (action implementation) offices and to regional agencies which must prepare, evaluate, and implement strategies to improve AEQ. In this activity the approach is toward integrated analysis of all forms of residuals. This is necessary to ensure that the mandated planning efforts under separate laws for different media do not result in inconsistent strategies.

4. The mismatch between political and administrative units and natural systems is widely recognized. Two approaches are possible. When institutional boundaries (political jurisdictions) are given and assumed fixed, the outputs of regional REQM modelling must be provided in a format that is understandable by, and useful to, the existing institutions. But regional REQM modelling also should generate information about possible desirable changes in institutional structures as parts of optimal REQM strategies.

5. Although a link between analysis and implementation authority is necessary if in fact any implementation is to occur, such close linkages can inhibit the direction, extent, and long-run usefulness of research on REQM and REQM modelling. Development of analytical methodology is not likely to be possible in the context of developing information for direct action decisions in the short run.

6. Researchers cannot initiate new methodological approaches and obtain immediate acceptance of both methods and their outputs by REQM agencies.

7. For REQM analyses to be useful, the user groups need to have certain attributes:

   a. ability to define problem area(s) of concern in a clear-cut way;

   b. capability of monitoring ongoing research and application studies;

   c. capability of continual updating and revision of the developed regional model, and its use; and

   d. internal staff to provide the capability specified above, rather than depend on outside consultants.

8. Regional REQM modelling appears to be at the stage where the models can be transferred to REQM agencies which have the resources and responsibility for continuing to develop and apply them to "real world" situations.

9. The forms of models can be transferred from region to region, but the empirical coefficients, transfer functions, and cost functions must be generated or validated for each region. For some types of activities, such as municipal sewage "treatment" plants, generalized cost functions may be sufficiently accurate for REQM modelling in many regions.

Concluding Observations

The following conclusions received more or less general support at the Rotterdam meeting. No attempt has been made to list them in order of importance.

1. It is highly desirable to develop regional REQM modelling to deal with or include acute toxics and chronic toxics. To do so probably requires simpler models in which time enters explicitly.

2. It would be useful to develop a handbook of gross residuals generation by type of activity. In doing so, there must be careful specification of the assumptions concerning: factor input prices; type(s) of raw materials; technology of production processes and materials recovery/by-product production/in-plant water and heat recirculation; product output specifications; socioeconomic characteristics of inhabitants, etc., associated with each coefficient. It would also be useful to develop a handbook of generalized cost functions for certain types of residuals modification processes.

3. Formal regional REQM models are useful for organizing research and, accompanied by preliminary sensitivity analysis, are useful in allocating analytical resources available in contexts where specific REQM strategies are to be developed.

4. The current lack of "benefit/damage" functions should not delay the application of regional REQM modelling. Analyses clearly can be made of the costs to achieve different levels of AEQ or different degrees of reduction in residuals discharges. Such analyses can also suggest:

a. where the "holes" in data are;

b. what the costs of filling the "holes," i.e., costs of obtaining information, might be;

c. what kinds of "cuts" might be made in order to produce outputs from the modelling; and

d. what the lossss are from not having information.

5. The degree of sophistication of a regional REQM model should be no greater than necessary to achieve the objective(s) of the analysis. Similarly, no one element in the model should be more refined than justified by

its relative importance and the accuracies of the other elements.

6. Little attention has been given to the use of regional REQM models for day-to-day operation of existing systems. How useful such models would be and what their characteristics should be, require investigation.

7. Better linkages are needed between regional REQM modellers and those collecting "environmental statistics," with respect to both economic activities and "natural world" systems.

8. More input into regional REQM modelling should be obtained from the public health professions.

9. The objective of regional REQM modelling is not to substitute for, or "second guess," the political process, but rather to provide information to that process concerning alternative REQM strategies, their costs and effects.